THE EVOKED VOCAL RESPONSE OF THE BULLFROG

A Study of Communication by Sound

Nelson Yuan-sheng Kiang

THE EVOKED VOCAL RESPONSE OF THE BULLFROG

A Study of Communication by Sound

ROBERT R. CAPRANICA

RESEARCH MONOGRAPH NO. 33
THE M.I.T. PRESS, CAMBRIDGE, MASSACHUSETTS

ACKNOWLEDGMENT

This is Special Technical Report Number 12 of the Research Laboratory of Electronics of the Massachusetts Institute of Technology.

The Research Laboratory of Electronics is an interdepartmental laboratory in which faculty members and graduate students from numerous academic departments conduct research.

The research reported in this document was made possible in part by support extended the Massachusetts Institute of Technology, Research Laboratory of Electronics, by the JOINT SERVICES ELECTRONICS PROGRAMS (U.S. Army, U.S. Navy, and U.S. Air Force) under Contract No. DA36-039-AMC-03200(E); additional support was received from the National Science Foundation (Grant GP-2495), the National Institutes of Health (Grant MH-04737-04), and the National Aeronautics and Space Administration (Grant NsG-946).

Library of Congress Catalog Card Number: 65-22004

Printed in the United States of America

FOREWORD

There has long been a need in science and engineering for systematic publication of research studies larger in scope than a journal article but less ambitious than a finished book. Much valuable work of this kind is now published only in a semiprivate way, perhaps as a laboratory report, and so may not find its proper place in the literature of the field. The present contribution is the thirty-third of the M. I. T. Press Research Monographs, which we hope will make selected timely and important research studies readily accessible to libraries and to the independent worker.

J. A. Stratton

PREFACE

Field studies of anurans in their native habitats have revealed that most species produce distinct vocalizations which probably serve species-specific functions. Recent anatomical and physiological studies have supported the view that the anuran possesses the apparatus and related nervous mechanisms necessary for sensitive sound communication. Unfortunately, because of certain peculiarities which seem inherent in studying their behavioral performance, we have little experimental knowledge of the sensory capabilities of these animals.

The work to be reported in this monograph originated in an attempt to bridge the existing gap between the naturalistic observations of sound communication in anurans and the anatomical and physiological findings in their auditory nervous system. The motivation for this research concerns my interest in the neural representation of complex meaningful sounds — stimuli for which the nervous system is presumably designed to process. Such a representation, although suggested by electrophysiological recordings of patterns of activity in various neural centers, requires a behavioral confirmation that these, in fact, are the principles by which information is processed within the intact animal. To this end, vocal behavior has been evoked from the males of laboratory colonies of bullfrogs (*Rana catesbeiana*) in response to a restricted class of natural and synthetic sounds. The evoked vocal responses, having the signal characteristics of the bullfrog's mating call, were highly selective. To playbacks of the mating calls of 34 different species of frogs and toads, only the mating calls of certain male bullfrogs would evoke calling from the laboratory animals. By employing a standardized experimental technique, quantitative measures of evoked calling have been obtained for a large number of synthesized bullfrog mating calls. This technique has permitted a detailed investigation of those features of an acoustic stimulus necessary to evoke the vocal response of the male bullfrog. These results have been summarized in a proposed model whereby the degree of calling to a given sound can be predicted. The model has been successfully tested with a number of natural bullfrog mating calls. In order that the reader might gain more insight into the relationship of the experimental results to recent naturalistic and physiological studies, I have attempted to include appropriate background material wherever necessary.

The research reported in this monograph was conducted while the author was a member of the Research Laboratory of Electronics at the Massachusetts Institute of Technology, and was the basis for a doctoral thesis submitted to the Department of Electrical Engineering in January, 1965. I wish to express my appreciation to Professor M. H. Goldstein, Jr., for his supervision and encouragement of this research. It was through my association with Professor Goldstein and Drs. L. S. Frishkopf and W. A. van Bergeijk that I initially became interested in the study of sound communication in the anuran. Professor K. N. Stevens has provided invaluable assistance in the synthesis of the acoustic stimuli that have been employed in this research. I also wish to thank Mr. M. J. Murray for his aid in the instrumentation of the laboratory and in helping to analyze and synthesize the bullfrog's mating call. The Communications Biophysics Group, under the guidance of Professors W. A. Rosenblith and W. M. Siebert, has been most cooperative in providing the diversified facilities needed to pursue this experimental research.

The support of the Bell Telephone Laboratories throughout my doctoral program is gratefully acknowledged. Their award of a Communications Development Training Fellowship made it possible for me to continue my graduate studies toward a doctoral degree.

Murray Hill, New Jersey
February, 1965

Robert R. Capranica

CONTENTS

PART I INTRODUCTION

PART II THE EVOKED VOCAL RESPONSE (EVR)

PART I

INTRODUCTION

Chapter 1

AIMS AND SCOPE OF THE RESEARCH

The main purpose of this research has been to develop and employ a behavioral method for an investigation of the ability of the anuran (frogs and toads) to discriminate among certain natural and natural-like sounds. The incentive for this investigation has been twofold. First, field observations by naturalists have suggested that sound communication probably plays a significant role in the lives of most anurans and that most species can apparently discriminate certain sounds made by members of their own species from those made by members of other species. Second, recent neurophysiological findings have provided detailed information about the coding of acoustic stimuli in the peripheral auditory system of these animals. The hope has been that our research might provide experimental evidence for the naturalist's hypothesis concerning the species specificity of anuran vocalizations and, at the same time, might relate the auditory capabilities of the anuran to the coding of these sounds in his nervous system. Since the bulk of the neurophysiological findings pertained to the bullfrog (Rana catesbeiana), this species was chosen for study.

In undertaking this research, we sought a behavioral method that would afford a quantitative measure of the bullfrog's ability to discriminate among auditory stimuli. In our judgment, a suitable method had not been adequately demonstrated by previous experimenters. Whereas classically and instrumentally conditioned reflexes have traditionally provided the means for studying auditory performance in other animals, the development of a conditioned reflex in the anuran in response to sound has not been so successful: attempts with different anuran species either have ended in failure or else have led to inconclusive results (despite the fact that these attempts have been both diversified and ingenious). Therefore, in view of the difficulties that had been encountered by previous experimenters, we considered several alternative methods in our preliminary studies.

An attempt was made to condition the heart rate of the bullfrog to sound. We chose to monitor cardiac responses because: (1) such a study with the anuran had not been previously attempted, and (2) the cardiac component of the conditioned reflex had been successfully employed in studying the sensory performance of other animals (e.g., fish,[1-3] dogs,[4-6] humans[7-10]). In our attempt the cardiac rate in unrestrained bullfrogs was monitored by means of electrodes implanted subcutaneously in the nasal region. Electric shock, applied through these same electrodes, was used as an unconditioned stimulus; sound (anuran mating calls, tones, pulse trains, noise bursts) was used as a conditioned stimulus. Changes in heart rate were obtained during experimental trials, but they were found to be unstable and non-repeatable. Inasmuch as relatively stable, reproducible responses are necessary for a detailed investigation of sensory performance, this attempt was abandoned. Furthermore, since previous experimenters had encountered similar difficulties in conditioning behavioral responses of the anuran, we decided not to explore other types of conditioned responses for the purposes of our research with the bullfrog.

A quite different method, involving the approach response of the female anuran to the male's mating call, was also considered. Martof and Thompson[11] had previously observed that female chorus frogs (*Pseudacris nigrita*), upon capture and separation from their mates during the height of the mating season, would often respond in the laboratory by approaching a loudspeaker through which mating calls of male chorus frogs were presented. This response has since been employed by several experimenters[12-17] to test the discriminatory ability of various species of female tree frogs (*Pseudacris*, *Hyla*). It has been found that, in many instances, the females will respond only to the mating calls of their own species.

Unfortunately, a study involving the approach response of the female entails a very serious limitation: females will respond only for a very short time while at their annual reproductive peak. Preovulatory or postovulatory females will not respond in the laboratory to the male's mating call.[12,17] The transient nature of the female's approach response makes it impractical for a detailed, quantitative study of mating call discrimination by the female. To answer questions of the precise interrelationships between call parameters that render discrimination possible would require a tremendous battery of simultaneous experiments. That is, since all of the females of a given species normally reach a reproductive peak at approximately the same time, experimentation is constrained to the brief period of the female's ovulation. An exceptionally large number of gravid females must therefore be involved during this brief period in order to reveal the basis for the female's discrimination. For this reason, we decided that a method involving the approach response of the female would be unsatisfactory for our research with the bullfrog. We might point out that one way of overcoming the difficulties encountered in a detailed study of the female's approach response would be to induce the sexual cycle artificially in laboratory females by hormonal injection. This possibility was considered but not pursued.

Another method, which we shall refer to as the "evoked-calling method," was also explored. This method involves a highly selective vocal response to an acoustic stimulus. We had observed, in maintaining a laboratory colony of several adult male and female bullfrogs, that spontaneous vocalizations were seldom made by these animals. However, occasionally a male of the colony happened to utter a distinct call having the signal characteristics of the bullfrog's mating call. This vocalization often was immediately answered by a similar call from a different male. It therefore occurred to us that it might be possible to evoke such a vocal response from the male bullfrog by presenting suitable acoustic stimuli. We investigated this possibility in several preliminary studies.

We found that the playback of a recorded bullfrog mating chorus evoked a high level of calling from the laboratory males. Their vocal responses, each having the properties of the bullfrog's mating call, were very selective. In presentations of the mating calls of 34 different species of frogs and toads, the males only responded to bullfrog calls. Further study indicated that the male's vocal response could also be selectively evoked by synthesized bullfrog mating calls: certain of these synthetic sounds were very effective in evoking calling whereas others evoked little or no calling. The evoked responses were stable and repeatable. We therefore decided that the evoked vocal response would allow a quantitative study of the male's ability to discriminate among certain natural and synthetic sounds. This response has been studied in detail in our research with the bullfrog. Our main concern has been to determine those properties that a sound must have in order to evoke the vocal response from the male bullfrog.

Evoked-calling experiments, involving different classes of sounds, have been conducted with two laboratory colonies of bullfrogs. The results of these experiments are summarized in a proposed model. The model allows prediction of the degree of evoked calling to a given sound on the basis of its spectral structure and temporal periodicity. The remainder of this monograph is devoted to a detailed description of the experiments and the formulation of the model. A summary of anuran sound communication precedes this description in order to provide suitable background material. Included in the discussion of the experimental results and the proposed model is the relevance of recent electrophysiological studies. The close relationship between the highly selective vocal response of the bullfrog to sound and the electrophysiological findings in the auditory system of this animal is one of the outstanding features of this research.

Chapter 2

SIGNIFICANCE OF ANURAN VOCALIZATION

2.1 Classification of Amphibia

The vertebrate phylum of the animal kingdom consists of five classes: Pisces (fishes), Amphibia (amphibians), Reptilia (reptiles), Aves (birds), and Mammalia (mammals). The fish is an aquatic animal, whereas most reptiles, birds, and mammals are terrestrial. The amphibian, as its name implies, is partly aquatic and partly terrestrial: most species possess the unique ability to breathe either under water or on land.[18] Fossil evidence indicates that many of the early amphibian forms have disappeared,[19] but the present-day amphibian still represents that evolutionary midstage as life emerged from the water.[20] Today there remain but three orders of Amphibia: Gymnophiona (caecilians), Caudata (salamanders), and Anura (frogs and toads).

Caecilians are subterranean, limbless, blind creatures that are found only in Central and South America, Southern Asia, and parts of Africa. (None occur in the United States or Canada.[21]) There are 75 different species living today.[22] They resemble overgrown earthworms in appearance, and range in adult length from the largest of $4\frac{1}{2}$ feet to the smallest of only 7 inches. The greatest body diameter reported for any species is slightly over one inch. Although little is known of the sensory capabilities of these animals, most are believed mute, and all lack any external sign of a specialized auditory receptor.

Salamanders, comprising 225 known species and subspecies, are found in all parts of the world except southern South America, most of Africa, and all of Australia.[23] They are scaleless, tailed, and lizard-like in form: four legs of equal length are used for movement on land – the tail is used in swimming. Species range in length from a few inches to over five feet (Megalobatrachus japonicus). Salamanders possess no vocal cords and are believed mute (except for the ability to emit a faint squeak when handled). As in caecilians, they lack a tympanum or other external sign of an auditory apparatus. Nevertheless, some species have been shown to be responsive to low-frequency sounds.[24] It has been suggested that hearing in the salamander occurs via bone conduction through the front legs and shoulder girdle to the inner ear.[25]

Anurans (frogs and toads) form the largest of the three amphibian orders. There are over 2600 known species and subspecies, ranging in body length from less than $\frac{1}{2}$ inch to over 12 inches.[26] They are found throughout the world in diverse and, sometimes, rather bizarre forms. Whereas caecilians and salamanders are believed voiceless and incapable of detecting faint sounds, most frogs and toads possess a distinct voice and an auditory receptor sufficient for sensitive hearing. As a consequence, aspects of sound communication have been attributed to most species. In view of the relevance of these aspects to this research, the more commonly accepted interpretations of anuran vocalizations will next be discussed.

2.2 Classification of Anuran Calls

The appearance of a vocal apparatus in the anuran marks the beginning of laryngeal vocalization in the vertebrate phylum.[27,28] The significance of the anuran's sounds has been, and continues to be, a source of controversy and speculation since at least the time of the early Greek philosophers.[29,30] That this controversy should persist today is, in part, a result of the vast folklore and superstition that have surrounded these animals. In some parts of the world, certain frogs and toads are still considered to possess supernatural powers.[31] Various African tribes hold magical rites involving the femur of Rana goliath (the largest living frog). Dried toads and their skins have been used for medicine by the Chinese for several centuries. And in southern China, the flying frog (Rhacophorus dennysi) is worshipped as a god and carried about in a chair! Such superstitious beliefs as these have led to widespread misconceptions of anuran vocal function.

The facts are that most anuran species utter distinct sounds and seem to possess an organized, though small, vocabulary. Naturalists believe that each call conveys a specific biological significance. Their field observations have identified certain environmental conditions under which different sounds are normally produced. Unfortunately, the different types of calls, other than the mating call, have not been extensively studied. A more thorough investigation of these calls is obviously needed in order to extend our understanding of sound communication in the various species. In this regard, different types of calls have been recorded from laboratory bullfrogs in the course of our research, and are described in Appendix A.

The following general discussion of anuran sounds, unless otherwise noted, is due chiefly to Bogert in his recent and detailed review.[27] Since the mating call is of primary interest in our research, its discussion will be emphasized.

2.2.1 Mating Calls. Calling by the adult male signals the advent of the mating season in most species. The ability to utter this call is believed a sign of sexual maturity: juvenile males are mute. Typically, in the spring or early summer, the adult males leave their normal (predatory) territories and congregate at suitable breeding sites. The calls of the males already assembled at these sites seem to attract other males whose calling adds, in turn, to the chorus. Bogert reports that, in some species, individual chorusing males alternate in call with one another and give the impression that one male "answers" the other. Alternate calling by different males leads to a gradual increase until the chorus reaches its full intensity. After an interruption, the animals seldom start calling simultaneously. Instead, a gradual increase in chorus intensity begins anew as individual males alternate in call. There is also evidence that a social hierarchy of calling exists within the chorus. Holmes[18] observed that the chorusing of bullfrogs seemed to be under command of a "captain" male. When the captain began croaking, all others of the chorus followed; and when he stopped, the others also stopped. Calling order among chorusing males has been reported in a few other species as well. Bogert discounts such evidence as being inconclusive and, instead, suggests that anurans are not capable of a social order of behavior.

In a number of species, the aggregation and chorusing of the males precede the appearance of the females at the breeding sites. The female, presumably attracted by the mating calls of the males, responds by approaching the chorus. Markuze,[32] in capturing over 2600 specimens of Rana radibunda at different times of the year, found that the males migrated to the breeding places during the late spring in advance of the females. It was only after the males had congregated and chorused that females began to appear. During the height of the summer mating season, Markuze noted an equal ratio of the two sexes.

The female, upon reaching the mating chorus, is clasped in amplexus by one of the males, and mating follows. She normally remains silent throughout this breeding period.[33] At the conclusion of the mating season, the males gradually cease their chorusing and then return, as do the females, to their original territories. In most species, males and females mate but once during each annual mating season.[34,35]

2.2.2 Release Calls. Males seem unable to discriminate sexes: during the mating season they clasp both males and females.[36] If the animal clasped is another male, he usually struggles and then emits a characteristic call, whereupon he is released. In contrast, a receptive female is silent when clasped, and the male retains his hold. The female, with the male mounted on her back, then deposits her eggs, and the male simultaneously fertilizes them externally. Upon termination of oviposition, the male releases the female, presumably because she no longer has sufficient girth or firmness.[36,37]

If a nonreceptive female (pre- or postovulatory) chances into the breeding area, she likely will be clasped by an ardent male.

She achieves her release by issuing a call very similar to that of the clasped male. Thus it is that, during the mating cycle, females are receptive and silent; when the cycle is over, the female regains her voice to notify admiring males that their attentions are no longer desired.

2.2.3 Territorial Calls. Except during the mating season, individuals in most anuran species tend to lead solitary lives.[38] Territorial calls, made by both males and females, serve to maintain individual territories. In most species, intruders are vocally forewarned; if such determent is unheeded, an attack by the proprietor follows. Such attacks, as Test[39] has observed, result in one or the other emerging victorious. The victor then takes up the task of defending his territory against subsequent intruders. Within a given species, there may be more than one type of territorial call. This will be discussed further, in relation to observations of laboratory bullfrogs, in Appendix A.

2.2.4 Rain Calls. In certain instances, male frogs (most commonly Hyla) have been heard to call with the rise in humidity prior to a thunderstorm or after a light rain.[27] Such vocalizations – not associated with mating behavior but presumably with changes in moisture content of the air – are broadly classed as rain calls. In some species, the call may resemble the mating call of that species; in others, it may resemble the release call. Observers often have reported that, in making the rain call, the male may only partially inflate his vocal sacs. In contrast, the mating call is made with the vocal sacs fully inflated.

The exact function served by the rain call, if in fact there is such a function, is obscure. For some species, it may simply be that an increase in humidity produces more activity in the animals, and it is the activity itself that triggers calling. In maintaining two laboratory colonies of bullfrogs during the course of this research, we found that sprinkling the animals with water occasionally initiated calling. However, a great variety of other conditions (e.g., the sound of splashing water, feeding, turning the laboratory lights on or off, airplanes flying overhead) produced similar vocal behavior. Such calling could not be ascribed to any one stimulus or situation (other than a general external disturbance).

2.2.5 Distress Calls. The distress call is quite unlike any other within the anuran's repertoire. It varies from a low-pitched sound within some species to a shrill scream in others. In nature, the distress call is often made by anurans when under attack by a mortal enemy. In the laboratory, it can be evoked by severe handling or by application of various noxious stimuli. The most characteristic feature of the distress call is that it is made with the mouth wide open: all other calls are made with the mouth tightly closed. The call is made by both sexes.

The biological significance of the distress call is not clear.

Frogs in the immediate vicinity of a distress call often show no overt reaction. Yerkes[40] reported that he was unable to induce a motor response in a natural colony of green frogs (Rana clamitans) by sounds of any sort as long as he remained hidden. Furthermore, he noted that the distress call of an attacked member of that colony failed to produce an observable reaction in other members of the colony. Yet, when Yerkes made himself visible, all of the animals jumped into the water. If the distress call does serve a purpose, it might simply be to alert other members of the species to possible predators. Visual or vibrational information apparently predominates in triggering escape reactions.

2.2.6 Warning Calls. Frogs, just prior to plunging into the water from a nearby bank, often emit a short, grunting sound. Naturalists have associated such calls with a warning function for others in the vicinity to be on guard for potential danger. This hypothesis seems circumstantial.

Our laboratory bullfrogs have often emitted relatively loud, explosive "grunts." Quite frequently, the animals followed with a leap into one of the terrarium ponds. Just as frequently, these same sounds were made prior to a leap out of the ponds. A common occurrence was also immediately after a male had issued a mating call. Rather than functioning as a warning signal, such grunting sounds seemed no more than a sudden expulsion of air from the lungs. In the case of our animals taking relatively long jumps into and out of water, deflation of the lungs provided a decided jumping and landing advantage. In the situation involving the male's mating call, upon termination of the call, the lungs were often fully inflated (see Chapter 3). Since this is not the normal mode of lung volume, the lungs were partially deflated and a "grunt" occurred. The assignment of a warning function to the production of this sound, at least in the case of our laboratory observations, appears questionable.

2.3 The Mating Call as a Species Isolation Mechanism

The mating call is considered to be the most important sound in the normal life of the anuran.[27] It is believed to play a major role in the preservation of most species (i.e., minimize interspecific breeding and promote more efficient reproduction). The basis for this belief resides in the species specificity shown by the anuran's mating behavior and in the uniqueness of the mating call of the male of each species.

In some geographical areas, there may be as many as 12 to 14 different anuran species breeding simultaneously within the same pond.[27,41] Nevertheless, heterospecific mating is rare, and natural hybrids seldom occur. Blair[42] has proposed that the distinctive mating call of the male serves as an isolation mechanism that enables the female to detect calling males of her own species. Such a mechanism promotes a positive attraction only between males and females of the same species and, therefore, allows a mating isolation within each species. Blair further suggests that the selectivity of this mechanism is the result of an evolutionary interaction between different species:

> The mating call is an especially favorable characteristic of anurans for the investigation of speciational and general evolutionary dynamics in this group. This is true because, with few exceptions, the function of the call is limited to the attraction of a mate. The essentially voiceless females are attracted to the calls of the males which generally precede them to the breeding pools. It is to be expected, therefore, that evolution of the call would be influenced by two major factors: (1) the function of attracting a female of the same species; (2) the disadvantage population-wise of attracting a female of another species. It is to be expected that the first function would tend to promote uniformity of call through the species population and that the second would influence differentiation between species.

Numerous field observations have supported Blair's proposal that the mating call serves to isolate the species for mating. For example, Lowe[43] observed several instances of female spadefoot toads (*Scaphiopus*) choosing calling males of their own species from a mixed chorus of two closely related subspecies. He then captured over 200 clasped pairs of this same mixed chorus but never found any heterospecific pairs in amplexus. Studies by other observers (e.g., Martof and Thompson,[11] Bogert,[27] Smyth,[35] Noble[37]) have likewise reported instances of females selectively approaching calling males of their own species (*Bufo*, *Hyla*). Generally, the male will cease his calling and will turn to clasp the female as she approaches his immediate vicinity. However, in some cases, the male may be so intent on his calling that the female must touch him to make him aware of her presence.[35,37]

Additional evidence that the mating call might play a highly selective role in the anuran reproductive cycle is offered by the analyses of field recordings of various species. Such recordings, taken at the breeding sites of chorusing males, have recently been made possible by the advent of improved electronic techniques.[44] Comparisons of spectral and temporal properties reveal that the mating calls within each genus show a gross similarity which distinguishes them from the mating calls of the other genera. Within a given genus, however, finer differences between mating calls exist which, in most cases, uniquely characterize the call of each species. Blair[42] believes that the gross similarity in the mating calls within a genus is indicative of an evolutionary relationship and that the finer interspecific call differences are indicative of a dynamic interaction between the species of that genus. He suggests that an adequate geographical sampling of anuran mating calls might reveal the past migrations of the various species.

Littlejohn,[45] in studying the mating calls of seven species of Crinia in Australia, found that the calls of allopatric males showed little geographical variability. However, in those regions where two related species overlapped and shared common breeding territories, the mating calls of each diverged in such a manner as to accentuate their differentiation. Littlejohn therefore concluded:

> In extensive sympatric situations, calls of each species are well-differentiated with no overlap in ranges of variation of at least one call component, usually call duration or pulse repetition frequency. It is suggested that the component showing greatest differentiation is probably the critical factor by which female discrimination occurs.

Littlejohn compared the mating calls of Crinia insignifera on the mainland of Australia with those on a neighboring, isolated island. He found very little difference in these calls even though the two populations had probably been separated for about 8000 years. Thus, incidental divergence of the mating call is slow and cannot account for the differences between the calls of related sympatric species. Rather, such differences have resulted from the geographical interference of one species with another.

The hypothesis that anuran mating calls act as species identification signals for reproductive isolation is an intriguing one. It has evolutionary implications of animal sound communication that extend far beyond the scope of our research. Yet, it is still a hypothesis. Granted that the mating calls of various anuran species can differ in several dimensions, such as spectral composition, repetition and trill rate, and duration of call,[42,45-50] comparative studies of these sounds do not, by themselves, reveal the basis for discrimination by the different species. Such knowledge can be gained only through critically designed experimentation.

Chapter 3

ANURAN SOUND PRODUCTION

3.1 Vocal Apparatus

Except for a few peculiar examples, such as certain Pipidae[51] and Ascaphidae,[52,53] the majority of anurans show a similar vocal apparatus. As in most land vertebrates, the vocal system is intimately related to the pulmonary system.

In the anuran, the paired lungs communicate with the bucco-pharyngeal cavities through a rudimentary larynx.[54] The lungs, ovoid thin-walled sacs, are capable of great distention and may be readily inflated through the larynx. They remain inflated if the body is cut open, since the larynx remains closed under passive conditions; when the air is released from the lungs, they shrivel to an inconspicuous size.[55] The connective tissue layer of the lungs contains, in addition to blood and lymph vessels, numerous unstriated muscle cells which give the lungs considerable contractile power.

The larynx opens in front into the pharynx through a slitlike glottis and behind into the lungs through a pair of openings.[55] In most species, the larynx is attached to the related hyoid apparatus by muscle and connective tissue. The upper edges of the glottis are supplied by muscles through which the glottis may be opened or closed. Vocal cords in the larynx consist of a pair of thickened, elastic lips extending longitudinally across the larynx passageway. They may easily be seen from above by spreading the two sides of the glottis. The tension on the vocal cords can be altered by the action of the laryngeal muscles. In general, the larynx of the female is anatomically identical to that of the male, except it is smaller.[56]

The pharynx – a relatively short, straight passage – expands into the buccal cavity of the anuran.[57] The soft lining of the mouth and pharynx is highly vascular and provides an important respiratory function. The buccal cavity is of comparatively simple shape. It communicates with the external environment through the two nares at the tip of the upper jaw. The nares can be quickly opened or closed by the action of striated throat and jaw muscles. Since the mouth is normally kept tightly closed by the tonic contraction of specialized musculature serving principally the lower jaw, air intake and expulsion occur through the nares.[55]

The most distinctive features of the anuran's vocal apparatus are, typically, two small, bilateral openings in the floor of the mouth of the male and a related vocal sac. Liu[28] has studied the vocal-sac openings in many different males from all parts of the world. In certain species, he generally found two round openings; in other species, the openings were often narrow and slitlike. However, there appears to be some variation in the vocal-sac openings within any given species. Sometimes males of the same species showed two round openings, two slitlike openings, or a round opening on one side and a slitlike opening on the other. Occasionally, in the same specimen, one opening was much larger than the other.

The vocal sacs of the males are diverticula of the lining of the buccal cavity and can be greatly expanded by inflation with air through the vocal-sac openings. According to Liu, the sacs consist of elastic yellow fibers and connective tissue: externally they are covered with a layer of striated muscle fibers, and internally they are lined with flattened epithelial cells. The skin covering the vocal sacs may be modified to form folds, or it may appear as normal skin. Liu has classified the anuran vocal sacs into three principal types, in each of which the external skin may or may not be modified: (1) median subgular, (2) paired subgular, and (3) paired lateral.

The median subgular type consists of an unpaired, medial vocal sac beneath the throat. This single sac appears in the majority of species and is especially predominant in the more primitive forms.* Liu, in examining 136 species of Bufo, found that 128 had an unpaired, medial vocal sac. In a study of 89 Hyla, the vocal sac was median in 83 species. The median subgular sac is usually accompanied by bilateral, slitlike, vocal-sac openings.

The paired subgular type consists of two separate sacs lying beneath the throat. Each sac communicates with the buccal cavity through an individual vocal-sac opening. This type of vocal sac appears to be derived from the median type by division at the midline. It is more commonly found in the family Ranidae (e.g., Rana catesbeiana, Rana grylio, Rana septentrionalis). Most of the paired subgular vocal sacs are served by round vocal-sac openings.

The paired lateral type consists of two separate vocal sacs situated behind and below the junctures of the upper and lower jaw. It apparently has evolved from the paired subgular type, and is the most specialized. Vocal sacs of this type are very few: they are found principally in certain Ranidae (e.g., R. clamitans, R. pipiens, R. palustris, R. sylvatica). Most of these species show round vocal-sac openings (each opening serv-

*In some very primitive species, vocal sacs are completely lacking (e.g., Xenopus, Pipa, Ascaphus). These species are considered the most primitive.

ing an individual vocal sac).

The female anuran possesses neither a vocal sac nor vocal-sac openings.[55,57] Since the mating call is made only by the male, the vocal sacs are believed responsible for the ability of the male to produce this call.

3.2 Vocal Production

The dynamics of sound production in the anuran are poorly understood. As far as is known, all vocalizations except the distress call (Chapter 2) are made with the mouth tightly closed.[27] The mating call is believed to be made by the male with the nares closed as well (i.e., a closed system).[28,37] Whether the nares are completely closed in the production of other calls is not clear. Since the mating call is the main concern in our research, only its production will be considered.

Prior to generating a mating call, the male of most species fully inflates his lungs by intake of air through the nostrils and larynx; the glottis is opened to permit a free inspiration passage. Subsequent contraction of the lung musculature forces pulmonary air through the larynx and sets the vocal cords into vibration.[55,57] "Puffs" of air, expelled through the glottis at a rate corresponding to that of the vibrating vocal cords, excite the buccal cavity and connected vocal sacs. With the mouth and nares tightly closed, successive "puffs" produce increasing distention of the vocal sacs.[58] The termination of the male's croak is marked, in most cases, by collapse of the lungs and full expansion of the vocal sacs. The sacs are then deflated either by opening the nares and expelling the air or else by opening the glottis and shunting the air back to the lungs. In the latter case the lungs, again inflated, are ready for the production of another croak. In some species (e.g., the bullfrog – see Chapter 5), the mating call may be comprised of a sequence of identical croaks. In such calls, the intervals between consecutive croaks depend, in part, upon the time required to reinflate the lungs.[27]

The repetition rate of the temporal fine structure within the mating call is governed by the vibratory rate of the vocal cords. The spectral characteristics of an individual temporal cycle depend, at least in part, on (1) the waveform of volume velocity at the glottis, (2) the resonant properties of the buccal cavity, and (3) the resonant properties of the vocal sacs. The latter are commonly viewed as the dominant factor in determining the spectral composition of the male's call.[27,28] In addition, the elastance of the walls of the vocal sacs may play an important role. This is particularly suggested by the mating calls of some species (e.g., *Rana catesbeiana*) wherein the spectrum appears to remain relatively constant while the inflation of the vocal sacs changes greatly during the call production.

There are wide variations of sound production in the different species which seem to be associated with the distinctive mating calls of the males of these species. The preceding description is representative of most Ranidae. Many Hylidae produce their calls in much the same manner. A number of toads deviate in that, during the call production, small amounts of air may be rapidly shunted back and forth through the laryngeal chamber to produce trill pulsations.[27] This mode of production enables some species of Bufo to make very prolonged calls (over 53 seconds in Bufo cognatus).[47] Finally, an extreme variation can occur in certain primitive Discoglossidae. Zweifel[59] claims that the male Bombina variegata completely lacks a vocal sac and, instead, produces its mating call by forcing air from the buccal cavity back through the pharyngolaryngeal passage into the lungs.

Chapter 4

ANURAN HEARING

4.1 Auditory System of the Anuran

In view of the rather drastic changes endured by life as it evolved from the water, it is not surprising that one finds considerable variation in the auditory systems of anurans. For example, the more primitive species, such as members of the Pipidae described by Patterson,[60] show an auditory apparatus that is little advanced over that of the fish. In contrast, in the higher species, such as the family Ranidae, a well-developed middle and inner ear can be found.[61] It is to this latter family that the following discussions apply.

4.1.1 Peripheral Auditory System. The auditory-receptor organ of the bullfrog has recently been described in detail (van Bergeijk and Witschi,[62] Frishkopf and Goldstein,[63] Geisler et al.[64]). Briefly, the auditory apparatus consists of a large external tympanic membrane linked to the oval window of the otic capsule (inner ear) via three middle-ear bones (plectrum, columella, operculum). Within the capsule are contained eight sensory structures, two of which, the amphibian papilla and the basilar papilla, are believed to be auditory in function. Each papilla lies within an evagination of the endolymphatic system which is separated from the perilymphatic system by a thin contact membrane; the perilymphatic system is in direct communication with the oval window. Thus, both papillae are well situated for the sensitive reception of airborne sound. The morphology of the amphibian papilla is more complicated than that of the basilar papilla and suggests a more specialized role in the reception of acoustic stimuli.

In describing the comparative anatomy of the labyrinth and its innervation, de Burlet[65] has identified the basilar papilla as the precursor of the cochlea of higher vertebrates. The amphibian papilla seems to be unique to the Amphibia: it has not been found in other classes of vertebrates.

The amphibian and basilar papillae are innervated by fibers from the posterior branch of the eighth nerve.[65] These fibers, arising from a ganglion of bipolar cell bodies within the otic capsule, terminate centrally within the medulla.[63]

4.1.2 Central Auditory System. The central auditory system of the adult anuran has not been investigated in sufficient detail.

The assignment of an auditory role to various central nervous centers appears to be based almost entirely on anatomical comparisons with other vertebrates (primarily mammals). In presenting these results, therefore, we should be aware that they have yet to be verified by appropriate physiological studies.

Larsell[66] has made a histological study of the developement of the auditory system in larval and adult frogs (Rana catesbeiana, R. pipiens, Hyla regilla). He found that, while the peripheral lateral-line system of the tadpole undergoes degeneration during metamorphosis, the central nervous system simultaneously transfers its neural connections to the peripheral auditory and vestibular systems. That is, the lateral-line organs of the tadpole are innervated by the peripheral processes of unipolar cell bodies located in the ganglia of the seventh (facial) and tenth (vagus) cranial nerves. The central processes from the ganglia synapse with cells in a densely packed nucleus in the dorsal portion of the medulla (acousticolateral area). In the course of the tadpole's metamorphosis, the cell bodies of this nucleus gradually migrate in opposite directions to form eventually two distinct dorsal and ventral nuclei. In their migration, these cells lose their synaptic connections with the peripheral lateral-line fibers and, instead, send out dendritic processes to synapse with incoming fibers of the eighth nerve. At the same time, the cell bodies associated with the lateral line in the seventh and tenth nerve ganglia undergo degeneration, accompanied by atrophy of the lateral-line organs. The completion of metamorphosis is marked by the disappearance of the entire lateral-line system.

In the adult frog, according to Larsell, the dorsal nucleus in the medulla receives primary fibers only from the amphibian and basilar papillae and is, therefore, assumed to be auditory in function. The vestibular fibers in the eighth nerve terminate in the ventral nucleus. It is not clear whether this nucleus also receives auditory fibers. (If so, the number of such fibers is probably small.)

Kappers et al.[67] consider the dorsal nucleus in the medulla of the adult anuran to be the homologue of the mammalian cochlear nucleus. Papez[68] suggests this same identification.

Larsell claims that the superior olive makes its first appearance in the anuran: it is not present in salamanders or other lower vertebrates. Noble[25] and Kappers[67] share this same view. The superior olivary nuclei are believed to be derived phylogenetically from cells of the reticular formation.[66] Each nucleus appears to receive both crossed and uncrossed fibers from the dorsal and ventral medullary nuclei. In their decussation, the contralateral fibers represent an incipient trapezoid body.

The lateral bulbotectal tract is believed to carry third-order auditory fibers from the ipsilateral and contralateral superior olives as well as secondary crossed and uncrossed fibers from the medullary nuclei.[66] The ascending auditory fibers in this pathway terminate chiefly in the nucleus isthmi in the tegmentum and the torus semicircularis (corpus posticum) in the tectum. Larsell,[66] Kappers,[67] and Papez[68] consider the lateral bulbotectal tract in the anuran to be homologous to the mammalian lateral lemniscus.

The nucleus isthmi consists of a cortex of small and large cells surrounding a scattered cluster of cells and fibers.[66,67] The bilateral nuclei communicate via the isthmial commissure. The homologue of the nuclei isthmi in the mammalian auditory system has been controversial.[69] Le Gros Clark[70] has suggested that these nuclei should be considered as homologous to a differentiated portion of the lemniscal system (most probably the dorsal nuclei of the lateral lemniscus).

The tori semicirculares, also known as the corpus posticum or corpora posteriora, form a pair of ventricular eminences in the midbrain.[67,68] These nuclei are covered completely by the optic tectum. They consist mostly of small cells with a considerable number of larger cells scattered among them. The tori of the two sides are connected by a small band of fibers. Kappers[67] and Papez[68] believe these nuclei represent the precursor of the inferior colliculi of mammals.

The projection of fibers from the nuclei isthmi and tori semicirculares to higher auditory centers has not been well established.

4.2 Physiological Studies

From a comparative point of view, the peripheral and central auditory system of the anuran presents, at many points, primitive parallels to those found in higher vertebrates. Physiological investigations of the mammalian auditory system have commanded a great deal of attention.[71-73] In contrast, physiological investigations of the anuran's system have been few in number.

Van Bergeijk[74] constructed large-scale models of the basilar papilla of the frog. By using stroboscopic illumination, traveling waves and relative displacement maxima were observed on the model's tectorium in response to sinusoidal stimulation. The direction of wave travel and the location of displacement maxima varied systematically with changes in stimulation frequency. Van Bergeijk concluded that the basilar papilla is capable of performing a simple frequency analysis according to the place principle. However, recent electrophysiological results (Frishkopf and Goldstein[63]) do not support this conclusion.

Strother[75] has recorded electrical potentials from the inner ear of the anesthetized bullfrog in response to pure tones. He obtained responses to frequencies from a few cycles per second up to about 3000 to 4000 cps. Upon cutting the columella of the middle ear, the responses were drastically reduced. Strother's results suggest a microphonic potential analogous to the cochlear microphonic[72] generated in the cochlea of mammals.

The first successful recording of auditory acitivity in the eighth nerve of the anuran appears to be due to Adrian et al.[76] Their gross-electrode studies in decapitated Rana esculenta and R. temporaria suggested that these animals are insensitive to all but very loud sounds. In contrast, neural responses in the eighth nerve were found to be quite sensitive to vibration. They further reported that the frequency range of sounds initiating a synchronized discharge in the auditory nerve of these species varied significantly with temperature. At 10°C, the maximum frequency that would produce a synchronized neural response was about 175 cps; at 30°C, this maximum frequency was approximately twice as great.

Glekin and Erdman[77] recorded from single units in the eighth nerve of decapitated Rana radibunda. They reported three types of units: (1) some fibers showed spontaneous activity but did not respond to sound stimulation; (2) some fibers showed spontaneous activity and could also be excited by sound; and (3) some fibers showed no spontaneous activity but could be excited by sound. The highest frequency that would produce excitation was 730 cps.

Axelrod[78] found three classes of single units in the auditory nerve of intact Rana pipiens. One class of low-threshold, rapidly adapting fibers responded to frequencies in the 600 to 700 cps band. Repetitive sounds having energy in this band produced time-locked responses up to approximately 100/sec; for higher repetition rates the firing patterns broke up into submultiples. A second class of spontaneously active fibers fired to tones up to a certain frequency; above this critical frequency, tonal stimulation failed to produce excitation and, instead, inhibited spontaneous activity. A third type, designated as "seismic detectors," was extremely sensitive to vibrations but did not respond to sound.

Frishkopf and Goldstein[63] have recorded from single units in the eighth nerve of the anesthetized bullfrog in response to acoustic stimuli. Their results showed that some units are extremely sensitive (comparable to that of mammals). The upper limit of frequency to which responses could be obtained, even for high sound levels, extended to about 4000 cps. Two types of auditory units were found: (1) "simple" units with maximum sensitivity between 1000 and 2000 cps, and (2) "complex" units with maximum sensitivity between 200 and 700 cps. Within the complex class were units that responded not only to sound but also to vibration. Whereas simple units could not be inhibited by acoustic stimuli, all complex units could be inhibited by appropriate acoustic signals; 500 cps was the most effective frequency for inhibition of all complex units. Frishkopf and Geisler[79] have recently demonstrated conclusive evidence that the simple units derive from the basilar papilla and the complex units derive from the amphibian papilla.

Temporal coding was also investigated by Frishkopf and Goldstein.[63] The responses of units of both types were time-locked to pulse trains at rates of 50 pulses/sec and 100 pulses/sec, but at 200 pulses/sec this synchronization disappeared. Such "following" capabilities may play an important role in the coding of temporal information in the nervous system of the bullfrog. This will be discussed in more detail in Chapter 10.

Sachs[80] has recorded from primary auditory neurons in anesthetized green frogs (Rana clamitans) in response to sound. His results closely resemble those of Frishkopf and Goldstein[63] for the bullfrog. Sachs found two main populations of single units: (1) low-frequency units most sensitive between 150 and 450 cps, and (2) high-frequency units most sensitive between 700 and 1700 cps. The responses of the low-frequency units to an excitatory tone could be inhibited by the addition of a second tone of appropriate frequency: 700 cps was the most effective inhibitory frequency for all low-frequency units. The high-frequency units could not be inhibited by acoustic stimuli.

The early electrophysiological study of Adrian et al.[76] suggested extremely poor auditory sensitivity in the anuran. However, the more recent studies of Axelrod,[78] Frishkopf and Goldstein,[63] and Sachs[80] have indicated high sensitivities to airborne sounds. The discrepancy between the earlier study and these more recent studies appears to reside in the experimental state of the blood supply to the inner ear. Frishkopf and Goldstein[63] reported that sensitive responses in the auditory nerve of the bullfrog were dependent on an intact blood supply to the otic capsule. If this supply was damaged, auditory responses were very difficult to obtain. Inasmuch as the investigations of Adrian et al.[76] were conducted on decapitated animals, their findings of poor auditory sensitivity presumably were due to destruction of this blood supply. Since they found sensitive vestibular responses in their decapitated specimens, the vestibular organ evidently is not so dependent on blood supply.

Electrophysiological recordings from central auditory centers in the anuran have not been reported in any detail. This may reflect the earlier misconception that these animals possess relatively poor auditory sensitivity. It may also reflect the sparsity of appropriate anatomical descriptions of the higher auditory centers. In any event, future neurophysiological studies of the central nervous system are needed for a better understanding of anuran sound communication.

4.3 Behavioral Evidence of Hearing

Yerkes[81] appears to be the first to offer experimental evidence of hearing in anurans. In studying the green frog (Rana clamitans), he found that the withdrawal movement of the hindlegs of a suspended, blindfolded animal in response to a blow on the head (by a rubber hammer) was modified by the presence of sound. Yerkes claimed that sounds of frequency from 50 to 10,000 cps were effective in modifying this withdrawal response. The upper limit of 10,000 cps seems highly questionable, particularly in view of the rather uncontrolled acoustic stimuli that Yerkes employed (splashing of water, pistol explosions, tuning forks ranging from 100 to 1000 vibrations per second, electric bells with metal gongs and wooden gongs, sudden hammer blows, whistles, and a variety of sounds produced by the human vocal organs). Furthermore, Sachs[80] has recently found that primary neurons in the eighth nerve of the green frog do not respond to frequencies above 3000 cps (regardless of intensity).

Hinsche[82] observed that a defense reaction (Unken reflex[83]) could be initiated in most frogs and toads by either visual or tactile stimulation. He was unable to evoke this reaction with sounds. Several years later, Ferhat-Akat[24] repeated Hinsche's experiments with salamanders and was successful. He established that low-frequency sounds between 32 and 244 cps affected various species of salamanders.

Kuroda[84] observed small, irregular changes in the breathing rates of Rana tadpoles in response to bell sounds. He noted an increase in breathing rates to certain bells and a decrease in breathing rates to others. Unfortunately, Kuroda did not discuss the characteristics of his bell stimuli (other than to describe the physical dimensions of the bell and clapper).

Corbeille[85] has studied changes in the bullfrog's rate of buccopharyngeal respiration[57] in response to sound. By using an ingenious arrangement of levers attached to the bullfrog's throat, she claimed that tones from 100 to 8000 cps produced a change in the rhythm of the throat movements of her animals. The upper limit of 8000 cps does not agree with the electrophysiological findings of Strother[75] or of Frishkopf and Goldstein[63] mentioned earlier. These latter studies suggested that the bullfrog's auditory range extends only to about 4000 cps.

Bajandurow and Pegel[86] claimed to have conditioned both the buccopharyngeal respiration rate and a general increase in motor activity of Rana esculenta to acoustic and visual stimuli. They found that the presentation of a sound or a colored light would produce a change in respiration rate, accompanied by voluntary movement, if but two or three immediately preceding trials had been reinforced with electric shock. The amplitudes of the responses to auditory stimuli were much greater than to visual stimuli. The responses quickly extinguished (in the absence of electric shock) in just two or three trials. Attempts to develop a differential response to different sounds or colored lights were unsuccessful. Such negative evidence of discrimination, however, is questionable, for it is doubtful that Bajandurow and Pegel did, in fact, condition their animals. Rather, because the responses appeared so quickly after reinforcement with shock and then extinguished so rapidly in the absence of shock, it would seem as though they had observed nothing more than the phenomenon of sensitization.[87] This conclusion is further strengthened by their finding that the responses disappeared if they presented trials too slowly or waited until the next day.

Schwarz and Volkmer[88] have developed a clever technique for determining the auditory thresholds of Rana esculenta and R. temporaria. By poisoning their frogs with gamma-hexachlorcyclohexan, they found that a motor response ("quiver reflex") to the presentation of a visual stimulus could be modulated by the simultaneous presentation of a pure tone. For both species this opticmotoric reflex, sensitive to tones from 40 to 5000 cps, yielded a unimodal audiogram with a minimum at 600 cps and a corresponding threshold of approximately 0.16 dyne/cm^2 at that frequency. Furthermore, audiograms taken after a prolonged exposure (24 hours) to a loud (6.3 dynes/cm^2) 600-cps tone revealed a selective increase in threshold at 600 cps and a relatively smaller increase at 1000 cps. Such a selective loss resembles that observed in higher vertebrates under similar traumatic conditions. In fact, Schwarz and Volkmer went so far as to conclude:

> Das Hörorgan des Frosches verhält sich offenbar ganz ähnlich wie das des Menschen.
> (The auditory organ of the frog obviously behaves quite similarly to that of the human.)

Strother[89,90] made several unsuccessful attempts to condition a leg-flexion response in bullfrogs to auditory stimuli. He recently investigated the galvanic skin response to sound in seven bullfrogs.[90] Strother found that a change in the bullfrog's skin resistance occurred on the presentation of a pure tone. By using such a response, he obtained audiograms for male and female bullfrogs. Based on the presented data for only two animals, it is difficult to ascertain if there are any reliable differences in the audiograms of the two sexes. The audiogram for each sex shows a multimodal curve with a range of from 20 to about 4000 cps. The maximum sensitivity observed by Strother is in rough agreement with the results of Schwarz and Volkmer,[88] but the shapes of the audiograms in the two studies differ markedly.

A set of novel experiments involving the instinctive response of the anuran to the mating call has been contributed by naturalistic studies. Goldsmith[91] captured several male and female spadefoot toads (Scaphiopus hammondii), placed them in an open glass container, and then observed their behavior during the mating season as he carried them toward a chorus of males of the same species. Goldsmith noted that, when he was within several hundred yards of the chorus, the captive toads became excited and active. A refinement of this principle, previously described in Chapter 1, has been provided by Martof and Thompson's finding[11] that gravid female Pseudacris nigrita would respond in the laboratory by approaching a loudspeaker through which mating calls of the male were presented. This finding has led to a number of studies with different species of female tree frogs in an effort to demonstrate a species specificity in the female's approach response.

Littlejohn and Michaud[15,16] found that gravid female Pseudacris streckeri would approach the playback of the male's call of their own species but would not respond to the calls of P. clarki. These two sympatric species often breed simultaneously in the same ponds. Their mating calls differ considerably in duration, dominant frequency, and trill rate.

Littlejohn[14] observed that ripe female Pseudacris t. triseriata would more often approach a loudspeaker presenting mating calls of their own species or of P. nigrita than of P. t. feriarum. Littlejohn made spectrograms of these calls and found that the calls of P. t. triseriata and P. nigrita are similar in duration, trill rate, and dominant frequency, whereas those of P. t. feriarum differ significantly in each of these dimensions. It is noteworthy that P. t. triseriata and P. nigrita are allopatric species with no known geographical overlap (therefore, the two species are not normally forced to discriminate each other's calls), whereas P. t. triseriata and P. t. feriarum are believed to be sympatric species and to share common breeding territories.

Blair and Littlejohn[13] presented mating calls of Pseudacris streckeri and P. ornata to gravid female P. streckeri. The calls of these two species differ principally in their spectral characteristics (the dominant frequency averages about 500 cps higher in P. ornata). The female streckeri were attracted to the mating calls of male streckeri but not male ornata. Blair and Littlejohn proposed that this discrimination is based on the spectral differences between the calls of the two species.

Littlejohn, Fouquette, and Johnson[17] noted that two types of mating calls can be distinguished within the species Hyla versicolor: one type is fast-trilled, and the other type is slow-trilled. Both types of calls are present in mixed choruses. The mating calls of the two types are similar in all features except trill rate. Ripe females taken from a mixed chorus would respond either to the fast-trilled call or to the slow-trilled call, but not to both.

As was pointed out in Chapter 1, a major difficulty is encountered in studies involving the approach response of gravid females. One can only work with these animals for a very short period during their sexual peak. Littlejohn et al.[17] have found:

> From our experience in discrimination experiments, it appears that females will respond to the male call only when they are at a reproductive peak. Females not yet ready to ovulate or those that have already oviposited have never shown any response in experiments at this laboratory.

Nevertheless these studies, in spite of their short-term nature, have revealed an important finding: in several species, females have been shown to be capable of discriminating the mating calls of their own from those of other sympatric species. Such a finding is in keeping with the concept of the mating call as a species isolation mechanism.

PART II

THE EVOKED VOCAL RESPONSE (EVR)

Chapter 5

THE BULLFROG MATING CALL – ANALYSIS AND SYNTHESIS

In our research of the evoked vocal response of the male bullfrog, we have made extensive use of synthetic bullfrog mating calls. The choice to concentrate on synthetic stimuli, rather than on natural calls, has permitted greater flexibility in the investigation of those acoustic properties necessary to evoke the bullfrog's vocal response. That is, the ability to alter the properties of prerecorded mating calls is restricted to such gross operations as filtering, temporal chopping, playback reversal, and changes in the speed of playback. In such operations as these, the parameters of the natural calls cannot be easily varied systematically or independently. In contrast, the parameters of synthetic stimuli are under direct control of the experimenter and can be manipulated at will.

An analysis of the bullfrog's mating call laid the framework for our synthetic approximations. The results of this analysis and the techniques that we have employed in synthesizing these sounds are now summarized.

5.1 Analysis of the Mating Call

The mating call of the bullfrog is a loud, sonorous sound and is the cause for his common name.[21] Its carrying power in nature must be great. This call, as in the mating call of most anuran species, is uttered only by the male.

The bullfrog's mating call consists of a sequence of several croaks. An example, recorded from one of our laboratory males, is shown in Figure 5.1. The gross temporal properties of the mating calls made in our laboratory by different males or by the same male at different times showed considerable variation. The number of croaks issued in a single call could vary from 3 or 4 to as many as 14 or 15. Individual croaks usually lasted from 0.6 to 1.5 seconds, and intervals between them varied from 0.5 to 1.0 second. The rise and fall times of the temporal envelope of any given croak differed widely. In some calls (e.g., the call of Figure 5.1), the fall times of the croaks were much sharper than the rise times; in other calls, the rise times were much sharper; and in still other calls, the rise and fall times were approximately equal. Based on an analysis of many different calls, a symmetrical envelope with rise and fall times of 200 to

(a)

(b)

(c)

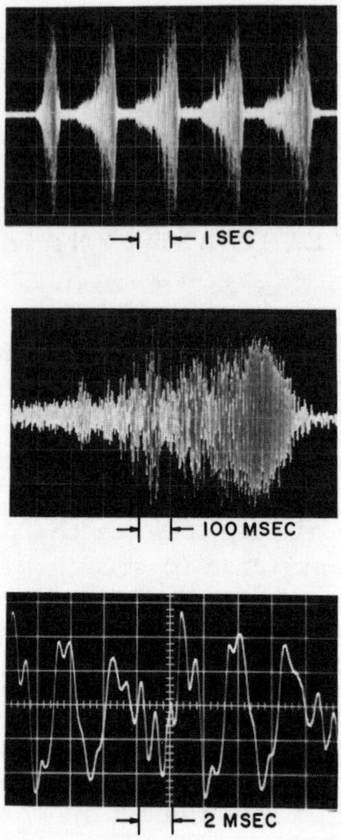

Figure 5.1. Mating call recorded from an adult male bullfrog of a laboratory colony. (a) The mating call, comprised of 5 croaks, in its entirety. (b) Second croak in the call, giving evidence of a pulse structure in the croak waveform. (c) Representative expansion of the second croak. Note that the temporal fine structure in the waveform has a repetition rate of approximately 100/sec; this periodicity is a distinctive feature of the croaks in the bullfrog's mating call.

300 milliseconds seems representative of an average croak within the bullfrog's mating call.

Though the gross temporal pattern of the bullfrog's mating call showed wide variations, the temporal fine structure was more stereotyped. A pulsatile, almost periodic waveform, having a repetition rate of approximately 100/sec, typified the fine structure throughout each croak of a given call. This structure is evident in the representative croak waveform for the mating call shown in Figure 5.1. The almost periodic nature of the temporal fine structure is one of the most distinctive features of the bullfrog's croak. The pulsatile character of the waveform is identifiable with the glottal excitation of the male's vocal sacs. Each fundamental period in the croak (Figure 5.1c), taken as the time interval between successive peaks in the waveform, represents the inverse of the instantaneous vibration rate of the vocal cords (see Chapter 3).

In a related study, Murray[92] has made a detailed investigation of the repetition rate within the bullfrog croak waveform. He was able to measure quantitatively the variations of successive periods in several croaks by displaying the waveforms on a high-speed pen recorder. Murray found that the instantaneous fundamental frequency within the bullfrog's croak (defined as the reciprocal of each fundamental period in the waveform) showed random, small variations. For a representative croak, the standard deviation of the instantaneous fundamental frequency was only 2.5% of the mean fundamental (95/sec). Murray concluded that the bullfrog's mating call can be properly described as a quasi-periodic sound.

The justification for classifying the call of Figure 5.1 (as well as similar calls from our laboratory males) as a mating call lies in several related findings. First, these calls have been made only by males. Females in our laboratory, though capable of producing other distinct vocalizations (see Appendix A), have never uttered a sound even remotely approximating that shown in Figure 5.1. Second, the generation of these calls by our laboratory males has always been accompanied by the full inflation of their vocal sacs. Field observations by naturalists have typified this act as characteristic of the male's production of the mating call. Third, acoustic presentations of these calls and of mating calls recorded in nature during the bullfrog's mating season have evoked similar vocal responses from our laboratory males. (The responses were of the same type as that of Figure 5.1.) And fourth, and perhaps the most significant, the signal characteristics of the laboratory calls were very similar to those of the natural mating calls.

Field recordings of the bullfrog's mating call were obtained

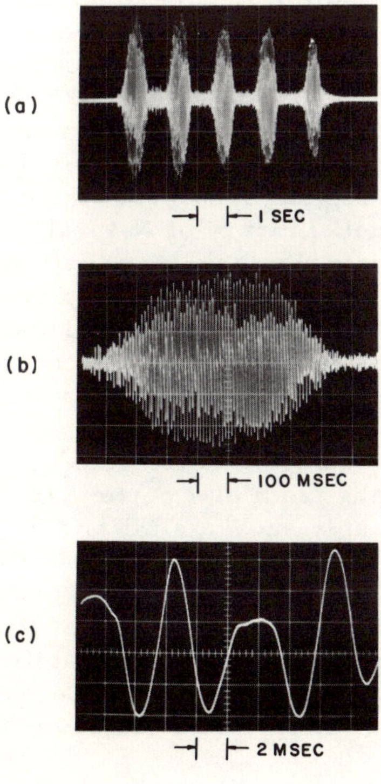

Figure 5.2. Field-recorded bullfrog mating call. This recording was obtained from C. M. Bogert (Department of Herpetology, American Museum of Natural History, New York City). (a) The mating call, comprised of 5 croaks, in its entirety. (b) Third croak in the call, clearly demonstrating the repetitive pulsatile nature of the croak waveform. (c) Representative expansion of the third croak, showing the waveshape in a single fundamental period. Note that the croak waveform has a periodicity of approximately 100/sec.

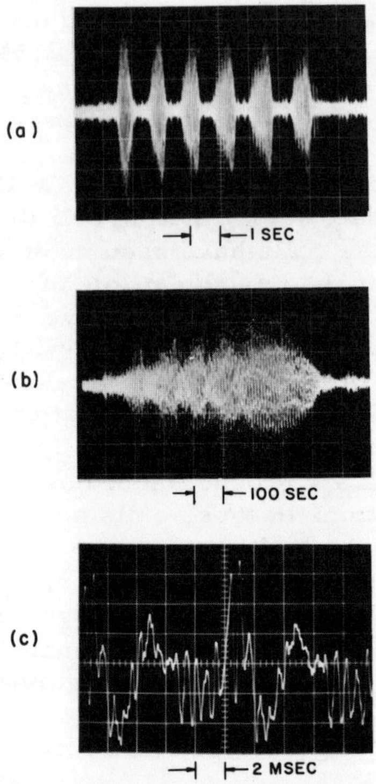

Figure 5. 3. Field-recorded bullfrog mating call. This recording was obtained from the Cornell Library of Natural Sounds (Laboratory of Ornithology, Cornell University, Ithaca, New York). (a) The mating call, comprised of 6 croaks, in its entirety. (b) Fourth croak in the call. (c) Representative expansion of the fourth croak, showing the waveshape in a single fundamental period. Note that the croak waveform has a periodicity of approximately 100/sec.

from C. M. Bogert (Department of Herpetology, American Museum of Natural History, New York City) and from the Cornell Library of Natural Sounds (Laboratory of Ornithology, Cornell University, Ithaca, New York). These recordings were taken at the breeding sites of bullfrogs during the height of the summer mating season. Two such mating calls are shown in Figures 5. 2 and 5. 3. The call of Figure 5. 2 was supplied by Bogert; the call of Figure 5. 3 was supplied by the Cornell Library of Natural Sounds. A comparison of these calls with the laboratory call of Figure 5. 1 shows a distinct correlation. Each call is comprised of a subset of individual croaks. The durations of the croaks are similar. Furthermore, the croak waveforms in the three calls are each characterized by a pulsatile structure with a periodicity of approximately 100/sec.

The waveshape in a single fundamental period of the croak waveform is representative of the croaks from that particular male: his croaks at other times invariably show a similar fine structure. In contrast, a comparison with the waveshapes in the croak waveforms from other males reveals a distinct difference between the individual animals (in spite of the fact that all of the croak waveforms show a pulsatile structure with approximately the same fundamental frequency – about 100/sec). As an example, consider the representative waveshapes in a single funda-

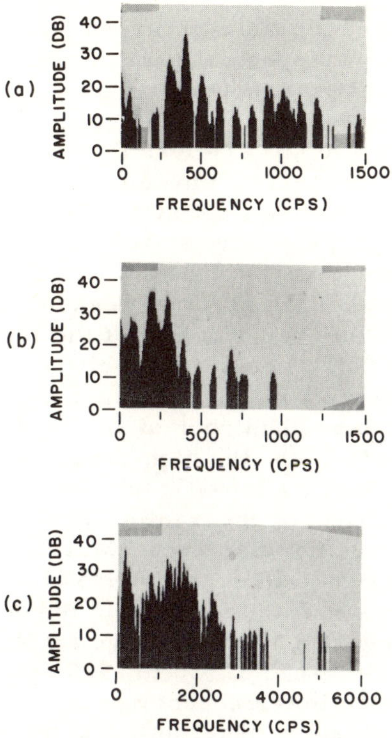

Figure 5.4. Spectral distribution of energy in the bullfrog's mating call. (a) A representative spectrographic section taken through the middle of the bullfrog croak of Figure 5.1b. (b) A representative spectrographic section taken through the middle of the bullfrog croak of Figure 5.2b. (c) A representative spectrographic section taken through the middle of the bullfrog croak of Figure 5.3b.

mental period of the mating calls of Figures 5.2 and 5.3. Although both calls show a fundamental of approximately 100/sec, the corresponding waveshapes clearly are different. The waveshape of Figure 5.2c is relatively simple and contains a predominance of low-frequency energy; the waveshape of Figure 5.3c is relatively complex and contains considerably more high-frequency energy. The waveshapes in the croak waveforms of other calls, for instance that of Figure 5.1, generally fell somewhere between these two in complexity.

Representative spectrographic sections[93] for the three mating calls of Figures 5.1 to 5.3 are shown in Figure 5.4. (All sections were made on a Kay Sona-Graph, Kay Electric Company.) Since the bullfrog's mating call is a quasi-periodic sound, sections taken at different points in a croak, as well as in different croaks of the same call, are similar.

The sections in Figure 5.4 emphasize the periodicity in the croak waveform by harmonic peaks spaced at multiples of the fundamental frequency (approximately 100/sec). A comparison of the sections indicates significant differences in the spectral distributions for the three calls. (These differences reflect the relative complexities of the waveshapes in the corresponding croak waveforms.) Nevertheless, certain common features can be recognized. First, all three calls show a major concentration of energy below 500 cps. In the case of the field-recorded calls (Figures 5.4b and 5.4c), a low-frequency spectral peak occurs between 200 and 300 cps; in the laboratory call (Figure 5.4a), this peak occurs between 300 and 400 cps. Second, the spectral amplitudes in the mid-range of 500 to 700 cps are less than that of the low-frequency peak. In fact, the spectral envelope of each call suggests a relative dip in the neighborhood of 500 to 700 cps. And third, high-frequency energy is present between 700 and 2000 cps, although its relative amplitude varies markedly in the three calls. The call of Figure 5.4c shows a spectral peak at approximately 1500 cps with an amplitude equal to that of the low-frequency peak. In the laboratory call (Figure 5.4a), a high-frequency contribution is evident between 800 and 1500 cps, but its spectral amplitudes are clearly less than that of the low-frequency peak. For the call of Figure 5.4b, the high-frequency components are barely noticeable. The Kay Sona-Graph has an amplitude threshold level when operated in the section mode: spectral amplitudes below this level are not displayed. Murray,[92] recognizing this fact, recently completed a more extensive study of the call supplied by Bogert (Figures 5.2 and 5.4b). He found higher-harmonic energy in this mating call. The average amplitudes of the spectral components in the range of 900 to 1500 cps were down by 25 to 30 db relative to the low-frequency peak.

In summary, the mating calls produced by different adult male bullfrogs generally have the following properties in common: (1) the mating call consists of a sequence of from 3 or 4 to as many as 14 or 15 croaks; (2) individual croaks usually last from 0.6 to 1.5 seconds; (3) silent intervals between successive croaks may vary from 0.5 to 1.0 second; (4) a symmetrical temporal envelope with rise and fall times of 200 to 300 milliseconds is representative of an average croak; (5) each croak is a quasi-periodic sound having a pulsatile waveform with a fundamental frequency of approximately 100/sec; (6) the spectral composition of each croak typically shows a distinct low-frequency peak below 500 cps, mid-frequency components (500 to 700 cps) lower in amplitude than that of the low-frequency peak, and high-frequency components (700 to 2000 cps) in various degrees of relative amplitude.

5.2 Synthesis of the Mating Call

The analysis and synthesis of vowels have been of major concern in studies of human speech. In general, vowels are quasi-periodic sounds[94] which can be uniquely characterized by their fundamental frequency and formant (i.e., relative energy maxima) structure.[95-99] Most vowels show four or more formants whose amplitudes and locations are determined by the shape of the vocal cavities.[100] (The amplitudes of the formants are not adjusted independently of their locations.[101]) Usually the first (i.e., lowest frequency) formant is the most intense.[101] In the production of a vowel sound, the glottis acts as a nearly constant source of quasi-periodic, sawtooth-shaped pulses of volume velocity, rich in harmonics.[102,103] The spectral structure comprising the formants therefore shows a harmonic relationship, with a fundamental frequency equal to that of the repetition rate of the glottal pulses.[100] Inasmuch as the repetition rate of the glottal pulses is roughly independent of formant locations, the time waveforms of most vowel sounds (produced by the same speaker at a given voice effort) have similar fundamental frequencies.[97,104,105]

The synthesis of vowel sounds has been realized principally through the use of electrical analog models of the human vocal system. Most models have taken the form of distributed- or lumped-parameter approximations of the transfer function of the vocal tract.[106] One such model, known as POVO[101,107] (Pole Analog of the Vocal Organs), has been developed by the M. I. T. Speech Communications Group for vowel production. POVO consists basically of four cascaded, simple, series-resonant circuits excited by a voltage pulse train. Both the resonant frequency and the bandwidth of each section are independently adjustable. The output from each successive resonant circuit is taken across the capacitive element. Consequently, POVO has four poles and no zeros in the finite complex-frequency plane. The output waveform is therefore a superposition of four damped sinusoids with frequencies corresponding to the first four formants. The periodicity in the temporal waveform is determined by the repetition rate of the pulse-train excitation. By employing a cascade connection of resonators, rather than a parallel combination, the relative amplitudes of the formants are adjusted naturally and automatically when the bandwidths of the formant resonators are maintained at appropriate values.[102] POVO has generated reasonably good approximations to vowel sounds. (E.g., Flanagan[101] investigated difference limens for vowel formant frequencies.)

POVO has also proved useful for the synthesis of bullfrog mating calls. Murray[92] recently was able to make close approximations to the temporal and spectral structure of the mating call of Figures 5.2 and 5.4b. He synthesized two types of sounds. In one type, the resonators were excited by a pulse train of constant pulse height and amplitude: the output waveform therefore had a periodic structure. In the other type, the fine-structure irregularities of the natural croaks were simulated by modulating the amplitude and repetition rate of the pulse-train excitation with white noise. Upon presenting these synthesized sounds to our laboratory bullfrogs, we found approximately equal calling to both types. We therefore concluded from this preliminary study that irregularities in the stimulus waveform would not be necessary in evoking the vocal response of the male bullfrog.

Our synthesis of the bullfrog's mating call has, for the most part, involved the use of POVO-type resonators (Figure 5.5) excited by periodic pulse trains (Hewlett Packard 212A pulse generator) or white noise (Grason Stadler 455B noise generator).

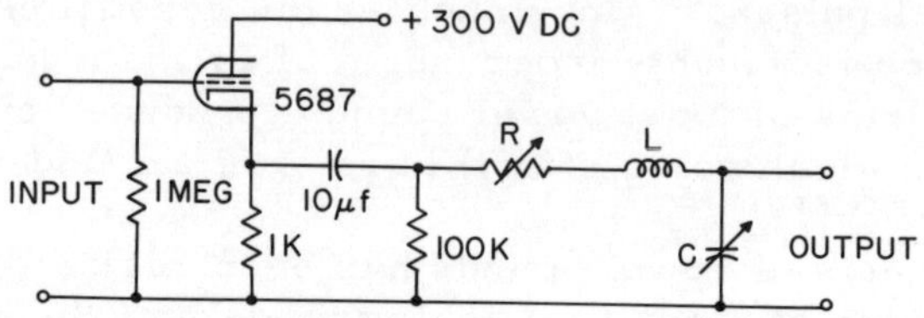

Figure 5.5. Circuit diagram of a POVO-type resonator. The resonant frequency and bandwidth of each resonator could be adjusted by the variable capacitance C (0.01-1.0 μf) and the potentiometer R (0-1000 ohms); the inductance L was fixed (0.5, 1.0, or 3.0 h, depending on the resonator). Four such resonators, connected in parallel, were employed in the synthesis of the bullfrog's mating croak (see Figure 5.6). The voltage impulse response for this parallel combination has four poles, with zeros interleaved, in the finite complex-frequency plane.

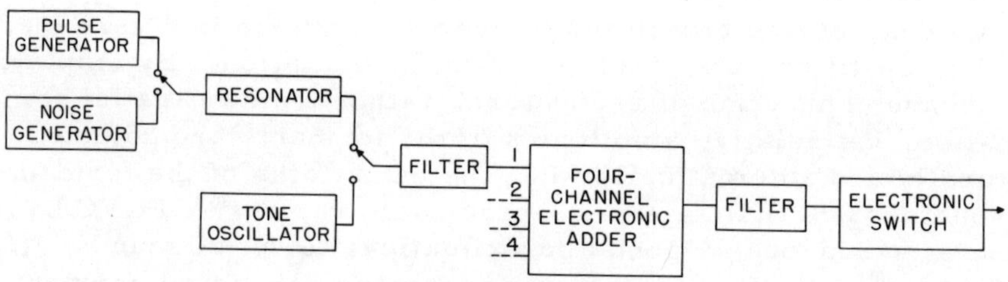

Figure 5.6. System diagram for synthesis of bullfrog mating croaks. Only one input channel to the adder is shown. (The other three were identical.) The filtered output of the adder was packaged into a sequence of identical croaks by means of the electronic switch.

However, instead of a cascade connection, a parallel combination of up to four resonant sections was chosen. Such a choice allowed an independent amplitude control of the contribution from each section. A four-channel, variable-gain electronic adder

(Argonaut LRA044) provided the parallel combination. Pure tone substitution (Krohn-Hite 430AB audio oscillator) for each of the resonant sections was also incorporated into the synthesis system. Electronic filters (SKL 302; Allison 2-BR), inserted between the resonator outputs and adder inputs, permitted independent filtering of each resonator output prior to combination. In addition, the adder output could be low-, high-, or band-pass filtered as desired. The filter configurations allowed greater flexibility in shaping the frequency spectrum of the synthetic stimuli. The filtered output of the adder was packaged, by means of an electronic switch (Grason Stadler 829B), into a series of bursts with duration, spacing, and rise-fall times similar to that of the natural croak. The synthesis system is represented by a block diagram in Figure 5.6.

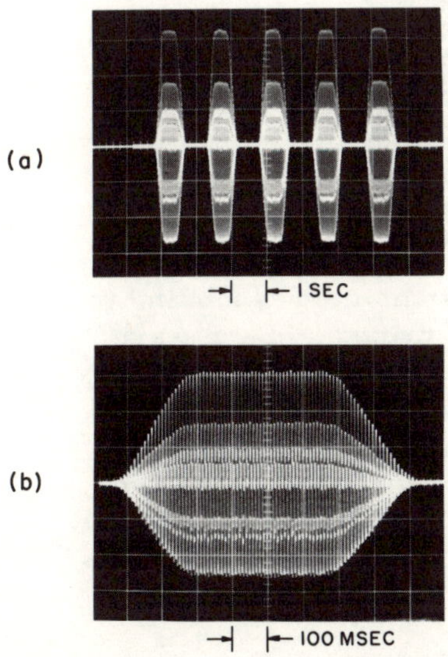

Figure 5.7. Temporal pattern of synthetic bullfrog mating calls. (a) A representative synthetic call comprised of five identically synthesized croaks. The number of croaks in any call were under the control of the experimenter. (b) A single croak from the above call. The waveform for this stimulus was generated by exciting the synthesis resonators with a periodic pulse train. The temporal envelope was made symmetrical, with rise and fall times of 250 milliseconds, by the electronic switch.

The temporal pattern of the synthetic mating calls was standardized by considering an average of several natural calls. The parameters adopted were (1) croak duration of 0.8 second, (2) croak rise and fall times of 250 milliseconds, and (3) silent intervals between croaks of 0.7 second. A sample synthetic call, consisting of five croaks, is shown in Figure 5.7. Note that the croaks repeat every 1.5 seconds and that the individual croak envelopes are symmetrical.

Three periodic waveforms, corresponding to the croaks of three representative calls synthesized by pulse-train excitation of the resonators, are shown in Figure 5.8. The periodicity in

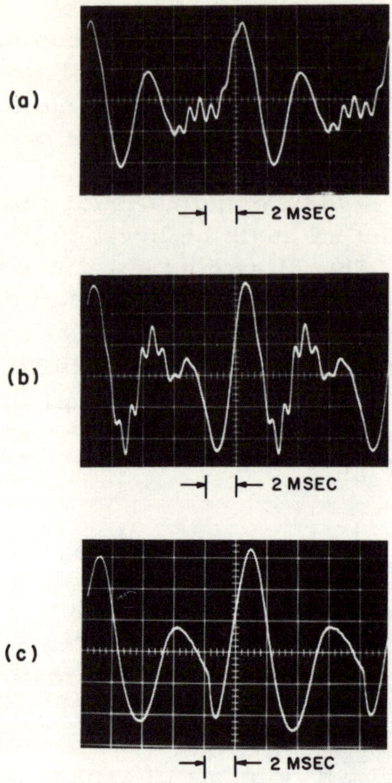

Figure 5.8. Three sample waveforms corresponding to three different synthesized mating croaks. Each waveform, having a periodicity of 100/sec, was generated by driving the synthesis resonators with a periodic voltage pulse train (repetition rate: 100 pulses/sec). The different waveshapes reflect the differences in their spectral energy (controlled by the bandwidths and resonant frequencies of the resonators, the relative voltage gains of the four-channel adder, and the filter settings).

each of the waveforms is 100 cps. Other periodicities were realized by varying the repetition rate of the pulse generator. A description of the different types of synthetic stimuli will be presented, in relation to their experimental results, in Chapter 9.

Chapter 6

THE EXPERIMENTAL SETUP

6.1 Ecological Aspects

Our studies of the evoked vocal response of the bullfrog have been conducted with two different colonies of bullfrogs kept in two specially designed terrariums. The terrariums have been maintained indoors in separate adjoining rooms in the M. I. T. Research Laboratory of Electronics. We have focused particular attention on the constancy of laboratory conditions in an effort to eliminate seasonal changes or cues. Individual air-conditioning units provided room temperatures of 67 to 72 °F during the entire two-year period of our research. An overhead incandescent lighting system automatically turned on at 7:00 AM and off at 8:00 PM (Eastern Standard Time) simulated day-night illumination levels. A dim lighting level was maintained in the absence of higher illumination during the evening and early morning hours.

Each terrarium was constructed of plywood in the form of a large rectangular box: one terrarium measured $5\frac{1}{2}$ ft (L) × 3 ft (W) × $2\frac{1}{2}$ ft (H), and the other terrarium measured 8 ft (L) × 2 ft (W) × 3 ft (H). Both were equipped with a hinged, screen top. A plastic liner, tightly fitted into each terrarium, provided a waterproof insert: the bottom sloped to one corner for drainage through an external outlet.

The floor of each terrarium was filled to a depth of 2 to 4 inches with gravel; a mixture of two parts of sand to one part of peat moss was laid over the gravel to an additional depth of 3 to 4 inches. A large pond, comprised of a plastic tank embedded in the layer of sand and peat moss, could be externally drained and refilled with fresh tap water by means of underground connective tubing. Large-mouth funnels, permanently inserted through the screened top, allowed live insects (mealworms, crickets) to be dropped into the terrariums for food for the animals. We therefore could water and feed our animals without significantly disturbing them.

One of the terrariums, housing a colony of adult bullfrogs and decorated with artificial plants, is shown in Figure 6.1. The plants, besides satisfying the aesthetic whims of the experimenter, provided a more natural surrounding and refuge — without them the animals choose to remain in the ponds. Live plants

Figure 6.1. One of the terrariums in which the laboratory bullfrogs were maintained.

were originally introduced but, because of lack of natural sunlight in the laboratory as well as abuse from the activity of the animals, they soon withered and were replaced. Also seen in Figure 6.1 are a loudspeaker mounted on the far wall of the terrarium for the presentation of acoustic stimuli and a microphone inserted through the screened roof for the monitoring of vocal responses from the colony.

Approximately 24 adult bullfrogs (18 males, 6 females) inhabited each terrarium. The members of one colony were originally captured in their native habitat in Wisconsin; the members of the other colony were captured in New Jersey. The Wisconsin animals, shown in Figure 6.1, ranged in body length from 6 to 8 inches. The New Jersey animals, somewhat younger, were $3\frac{1}{2}$ to 5 inches in length. There was no noticeable difference, other than size, between the Wisconsin and New Jersey frogs.

The animals within each terrarium colony were, of necessity, of the approximate same size, else the larger, by virtue of the confined quarters, would have quickly devoured their smaller "frogmates." The bullfrog has a voracious appetite and will attempt to swallow almost any sort of living creature — including other frogs — capable of being placed in his mouth. In fact, the bullfrog, the largest North American frog, is the scourge of the naturalist.[108] His widespread presence is a constant threat to the existence of smaller anuran species. However, the bullfrog is not without enemies of his own. Undoubtedly his worst nemesis is man disguised as a "frogleg gourmet."

Two of the New Jersey bullfrogs are shown in Figure 6.2. The animal on the left is a male; the one on the right is a female.

Figure 6.2. Adult male (left) and female (right) bullfrogs.

The external tympanum (eardrum) can clearly be seen as a round circular membrane directly behind the bulging eyeball. It is much larger in the male than in the female. The significance of this dimorphism, which also occurs in a few other species (mostly Rana), is not known.[21,27,109] The male in Figure 6.2 possesses paired subgular vocal sacs. They are not readily discernible except during his production of the mating call.

Since the anuran is commonly a host to a number of different types of microscopic and macroscopic parasites, he is susceptible to several highly contagious and often fatal diseases.[110,111] The bullfrog, by virtue of his more aquatic nature and indiscriminate choice of food, is particularly prone to such diseases. Consequently, the maintenance of our animals during the prolonged period of this research has posed a serious problem, especially in view of the relatively large number of animals within each terrarium. We have attempted to reduce the possibilities of disease by (1) maintaining the laboratory temperature between 67 and 72 °F (considered to be the approximate optimum temperature for several species of Rana: for example, R. pipiens[112]); (2) daily changes in the water of the terrarium ponds; (3) the addition of a slight amount of Acriflavine* to the water used in refilling the ponds; (4) immediate removal of any animal that showed signs of illness. Furthermore, since prolonged unnatural diets can produce severe bodily disorders,[110] the diets of our animals have consisted principally of live insects (although such food itself may introduce disease).

For the most part, we have succeeded in keeping our laboratory bullfrogs in a relatively healthy state during the period of our research. Occasionally we found animals that showed evidence of disease; e.g., lack of appetite and activity, hemorrhaging beneath the skin ("red leg"), edema, ulcerations of the skin or tympanum, abnormal discoloration of the skin. These animals, for fear of contaminating the others, were immediately removed from the terrarium. In several cases we were successful in curing the diseased animals (to the extent that they could be returned to the terrarium). A brief discussion of the treatment that we have employed will be found in Appendix B.

6.2 Acoustic Monitor and Presentation System

A dynamic omnidirectional microphone (University model 70) was vertically suspended through the screened roof of each terrarium (Figure 6.1). The microphone permitted a continuous monitor of background sounds and animal vocalizations. Its

---

*The amount of Acriflavine (Abbott Laboratories, North Chicago, Illinois) added was just sufficient to color the water a very pale yellow.

output, fed into a specially designed microphone preamplifier,* was amplified and transmitted via a coaxial cable to an audio amplifier (Magnecorder PT6) for listening at a remote location. Affiliated with the monitor system was a tape recorder (Ampex 601) for the recording of vocalizations from the terrarium animals. The over-all frequency response of the monitor and record network was flat to within ±4 db from 100 to 10,000 cps.

Integrated with the monitor system was a remote-controlled electronic system for the presentation of natural and synthetic sounds through the terrarium loudspeakers (KLH model 11). The acoustic stimuli were either generated from prerecorded magnetic tapes and associated tape recorders (Ampex 601; Bell T-347) or else synthesized by the synthesis system described in Chapter 5. In either case, all stimuli could be low-, high-, or band-pass filtered as desired. Each terrarium loudspeaker was driven via a coaxial cable by a variable-gain power amplifier (Magnecorder PT7). The acoustic level was controlled by adjustment of the amplifier gain. A select switch at the remote location permitted the experimenter's choice of either terrarium for study. The electronics comprising the entire presentation system (i.e., exclusive of the loudspeaker and terrarium acoustics) were flat to ±2 db from 100 to 10,000 cps.

6.3 Terrarium Acoustics

The terrarium acoustics were not ideal. Sound reflections from the terrarium surfaces and surrounding laboratory spaces introduced phase distortions and frequency resonances. In addition, extraneous laboratory noise produced a nonquiet background level.

The frequency response of the loudspeaker in each terrarium was measured, at the approximate height of the bullfrog's eardrum above the terrarium floors, with sensitive calibration equipment: Bruel and Kjaer 4131 condenser microphone, 2604 microphone amplifier, and 3304 automatic frequency response recorder. The floor of each terrarium was divided into a grid of 12-inch squares, and measurements were taken approximately in the center of each square. The results indicated that, except for a few location-dependent frequencies,† the frequency response in each square was flat to within ±5 db from 100 to 3000 cps (which is

*The preamplifier circuit was a modified version of the phonotape preamplifier, Figure 9.1, in the General Electric Transistor Manual (General Electric Company, Semiconductor Products Department, Syracuse, N.Y., 1962), Sixth Edition, pp. 125-128.

†At these few frequencies, the responses deviated by more than ±5 db. In some extreme cases, they deviated by as much as ±10 db.

the range of interest for the bullfrog). The peaks and dips in the frequency responses were very sharp and highly dependent on position: movement of the microphone by as little as 2 inches often drastically changed the magnitude of these relative maxima and minima.

From these measurements we concluded that, except for a few frequencies, the relative amplitude of each spectral component in a resultant acoustic stimulus would not deviate from the relative amplitude of the corresponding component in the electrical input waveform to the loudspeaker by more than ±5 db (over the range of 100 to 3000 cps) throughout each terrarium. This conclusion was verified by presenting several representative synthetic mating calls through the terrarium loudspeakers. We found that the spectral envelopes of these acoustic stimuli at different points in each terrarium generally did not deviate from that of their corresponding electrical inputs by more than ±5 db at any given frequency between 100 and 3000 cps.

Although the frequency responses in each terrarium were not ideal, they nevertheless were not so severe as to detract significantly from our experimental results – they simply were accepted and taken into consideration. They necessarily were accepted because of the requirement of animal confinement in a seminatural terrarium for the successful initiation of vocal behavior in the laboratory. Under these conditions, the realization of an anechoic terrarium, flat to frequencies as low as 100 cps, would be exceedingly difficult. They were taken into consideration by comparing only those synthetic mating calls whose relative spectral envelopes differed by a minimum of 10 db in the amplitude of at least one frequency component. This latter point should be borne in mind in our subsequent discussions of the experimental results.

Reverberation and phase distortion produced variations in acoustic temporal structure within each terrarium. A complex periodic signal into the loudspeaker resulted in a periodic acoustic waveform of the same fundamental as the input, but waveshapes recorded from different points in either terrarium differed considerably. Consequently, the temporal fine structure (other than periodicity) of the acoustic stimulus at the level of the frog's eardrum could not be uniquely specified and, therefore, was not investigated in our research. Experimental evidence, to be presented in Chapter 9, suggests that the exact waveshape within the bullfrog's mating croak is relatively unimportant – what seems to be more important is the periodicity within the croak waveform. However, a more precise demonstration, involving the acoustic preservation of temporal fine structure, is needed to verify this supposition.

Finally, the laboratory was not soundproof. An attempt to shield the terrariums from extraneous noise was effected, in part, by placing each terrarium in a separately enclosed room. In addition, the individual rooms were themselves housed in an enclosed laboratory. Such an arrangement tended to maintain a stable background noise level of 65 to 70 db SPL (re 0.0002 dyne/ cm^2, integrated over the range of 20 to 20,000 cps). Samples of the noise environment, taken at various intervals during our experimental research, yielded similar spectral characteristics. The spectrum of the background noise was not flat. A typical distribution, represented by an average of many samples, is shown in Figure 6.3 over the frequency range of interest for the bullfrog. As is often the case in room acoustics,[113] the relative amplitudes of the low-frequency components exceeded those of the high-frequency components.

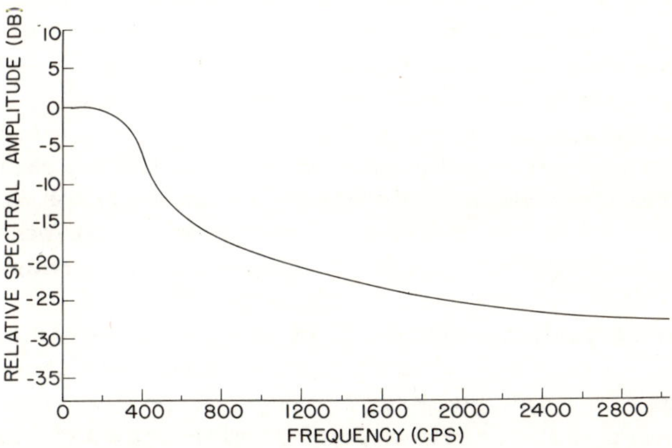

Figure 6.3. Spectral distribution of environmental noise in the laboratory. The envelope of the spectrum represents a smoothed average of many samples taken in the terrariums under typical experimental conditions (all amplitudes shown relative to that at 100 cps). In each sample, the amplitude at any given frequency between 100 and 3000 cps generally fell within ±10 db of the envelope value. (Measurements were taken with a Bruel and Kjaer 4131 condenser microphone, 2604 microphone amplifier, and 3304 automatic frequency response recorder. A continuous monitor of the noise distribution during actual experimental sessions was provided by analyzing the outputs of the terrarium microphones with a Hewlett Packard 302A wave analyzer – see text.)

In view of the nonquiet environment, precautions were taken to insure that the acoustic stimuli from the terrarium loudspeakers were not masked by room noise. All stimuli were monitored through the terrarium microphones and periodically analyzed with a Hewlett Packard 302A wave analyzer. The acoustic level of each stimulus was adjusted such that its spectral components, in the presence of the background noise, could be clearly detected with the wave analyzer. These levels varied, depending on the particular stimulus, from 64 to 76 db SPL. In addition, various representative stimuli were presented at several acoustic levels to insure that the performance of the terrarium animals was not dependent on the masking effects of the background noise.

6.4 Initiation of Vocal Behavior

The bullfrog is a wild, highly instinctive animal. His capture, transport, and then release within a laboratory terrarium (regardless of how "natural" one attempts to make it) represent a significant departure from his native habitat. Considerable time and effort are required to habituate these captive animals to terrarium confinement. The reward of such patience is the possibility of observing some rather interesting behavior.

Our bullfrogs, upon their initial release in the laboratory, tended to congregate in the four corners of the terrarium. The first several days were dedicated to repeated leaps against the walls or roof of the terrarium. Providing the animals were left essentially undisturbed, such jumping behavior gradually subsided.

The next stage in the habituation process involved a state of immobility. For most of the animals, this state lasted from several days to a few weeks. During this period, the animals either entered the terrarium pond or else crowded into the corners of the terrarium and then remained silent and motionless for prolonged periods of time. We were able to draw them slowly out of hiding by introducing food.

The frog has a relatively low metabolic rate (18 to 180 times slower than that of mammals, depending on environmental temperature[110]), and therefore needs not eat often. Food intake once every week is adequate. However, since food seemed to be a stimulus capable of initiating activity in our laboratory bullfrogs, feeding sessions were conducted either daily or else every other day. The amount of food offered in any given session was small so as to reduce the possibility of overfeeding. These early attempts involved the presentation of pieces of beef heart or liver on the end of a fine thread. Since the frog will not strike at a stationary target, the pieces of meat were dangled in a manner simulating a live insect. The thread, knotted at the end, was skewered through the beef heart or liver portion. If an animal chanced to take the food, the thread was released by a gentle tug.

The initial feeding attempts were relatively unsuccessful. The animals remained motionless in spite of efforts to place food strategically within their reach. However, after several trials, some of the animals began to orient toward the meat; the more aggressive members accepted it as food. Once feeding commenced, competition among neighboring animals incited others to join in. Within the following weeks, most of the colony were actively competing for food. The laboratory diet in later feeding sessions was supplemented with live insects (mealworms, crickets).

The rivalry between different members of the colony became rather keen; considerable fighting and territorial behavior gradually appeared. The more dominant animals tended to maintain choice terrarium locations. It was during this latter behavior, approximately 2 to 3 months after confinement, that the laboratory bullfrogs began to vocalize. In the months that followed, vocalizations continued and could be stereotyped into distinct types of calls. A description of these calls will be found in Appendix A.

Chapter 7

THE EVR TO NATURAL MATING CALLS

The spontaneous vocal behavior of our laboratory animals became most pronounced during and immediately after a feeding session. We noted that occasionally one of the males would utter a call which resembled the bullfrog's mating call. It often was immediately answered with a call of the same type from another male of that colony. This vocal behavior suggested that calling could perhaps be evoked from the terrarium males by the acoustic presentation of the mating call. A study followed in which the mating calls of several anuran species were tested for their ability to evoke a vocal response from our laboratory bullfrogs.

7.1 Methods

Field-recorded mating calls were obtained from several sources: (1) tape recordings supplied by W. F. Blair (Department of Zoology, University of Texas, Austin, Texas); (2) tape recordings supplied by C. M. Bogert (Department of Herpetology, American Museum of Natural History, New York City); (3) tape recordings supplied by the Cornell Library of Natural Sounds (Laboratory of Ornithology, Cornell University, Ithaca, New York); (4) phonograph record Voices of the Night by P. P. Kellogg and A. A. Allen (Cornell University Records, a division of the Cornell University Press, Ithaca, New York). From these recordings, excerpts of the mating calls of 34 anuran species were selected and rerecorded on magnetic tapes suitable for playback over the laboratory sound system. Each excerpt contained the isolated mating calls and composite mating choruses of one of the 34 species. Table 7.1 designates the species chosen in our study. This list includes representative members of the more populous genera of North America: Rana, Bufo, Hyla, Microhyla, Scaphiopus, Acris, and Pseudacris.

Experimental trials consisted of the playback of the excerpts through the terrarium loudspeakers and the simultaneous monitor of the vocal behavior of the laboratory animals. In all trials, the excerpts lasted for approximately 100 seconds. Each trial involved a monitor of background noise and spontaneous calling for a 5-minute prestimulus interval. If the noise level was stable and calling had not occurred, then an excerpt was presented. (The playback was omitted in the event that extraneous noise or calling

Table 7.1. Anuran Species Whose Mating Calls Were Tested for Their Ability to Evoke the Vocal Response of the Male Bullfrog

1. Bullfrog (Rana catesbeiana)
2. Southern Bullfrog (Rana grylio)
3. Green Frog (Rana clamitans)
4. Gopher Frog (Rana capito)
5. Meadow Frog (Rana pipiens pipiens)
6. Southern Meadow Frog (Rana pipiens sphenocephala)
7. Mink Frog (Rana septentrionalis)
8. Wood Frog (Rana sylvatica)
9. Pickerel Frog (Rana palustris)
10. Sphagnum Frog (Rana virgatipes)
11. American Toad (Bufo americanus americanus)
12. Hudson Bay Toad (Bufo americanus copei)
13. Southern Toad (Bufo terrestris)
14. Western Toad (Bufo cognatus)
15. Oak Toad (Bufo quercicus)
16. Fowler's Toad (Bufo woodhousii fowleri)
17. Spring Peeper (Hyla crucifier)
18. Common Tree Toad (Hyla versicolor)
19. Green Tree Frog (Hyla cinerea)
20. Anderson's Tree Frog (Hyla andersonii)
21. Barking Frog (Hyla gratiosa)
22. Squirrel Tree Frog (Hyla squirella)
23. Bird-voiced Tree Frog (Hyla avivoca)
24. Pine Woods Tree Frog (Hyla femoralis)
25. Giant Tree Frog (Hyla septentrionalis)
26. Narrow-mouthed Toad (Microhyla carolinensis)
27. Spadefoot (Scaphiopus holbrookii)
28. Western Spadefoot (Scaphiopus hammondii)
29. Cricket Frog (Acris gryllus)
30. Southern Swamp Cricket Frog (Pseudacris nigrita nigrita)
31. Eastern Swamp Cricket Frog (Pseudacris nigrita feriarum)
32. Least Swamp Cricket Frog (Pseudacris ocularis)
33. Western Swamp Cricket Frog (Pseudacris nigrita triseriata)
34. Ornate Chorus Frog (Pseudacris ornata)

chanced to occur during the prestimulus interval.) All vocalizations evoked* during the 100 seconds of playback, as well as the following poststimulus period of 80 seconds, were monitored and

*Our laboratory animals seldom called spontaneously except during feeding sessions. Upon termination of a feeding session, calling soon ceased and silence ensued for prolonged periods of time. Such behavior allowed a clear distinction between evoked and spontaneous calling.

simultaneously recorded on magnetic tape for subsequent analysis.

Experimental sessions were conducted daily over a period of several weeks. Each session generally lasted for four to eight hours. Within a session, the mating calls of the various species were presented in random order. Trial rates were randomly varied from once every ten minutes to once every hour. Each excerpt was tested a minimum of 10 to 20 trials at sound pressure levels ranging from 58 to 82 db re 0.0002 dyne/cm^2.

7.2 Results

Playbacks of the bullfrog excerpt were vociferously answered by the laboratory males: their vocal responses (see Figure 5.1 for an example) had the same signal characteristics as the field-recorded mating calls (e.g., Figures 5.2 and 5.3). The females of the terrarium colonies did not vocally respond in these trials.

Calling by the males invariably began within 10 to 45 seconds and then continued throughout the duration of the bullfrog excerpt. Upon termination of the playback, the males usually ceased their calling within 45 to 60 seconds and, thereafter, remained quiet until the next trial. A number of males participated in these evoked vocal responses: in some trials as many as 10 to 12 different calling males could be identified. Furthermore, a given male often called more than once during a single trial.

In remarkable contrast with the evoked vocal responses to the bullfrog excerpt, the playbacks of the mating calls of the other 33 species failed to initiate a calling response: our laboratory animals (both males and females) typically remained still and silent. We have repeated these trials at random intervals during the remaining $1\frac{1}{2}$ years of this research and have obtained similar results. In no instance could vocal behavior of any type be reliably evoked from the bullfrogs in either terrarium by the mating calls of species other than the bullfrog (regardless of trial rate or acoustic level).

The effects of trial rate on the degree of evoked calling to the bullfrog excerpt were briefly investigated. We repeatedly found that calling within an experimental session would gradually diminish if successive trials were presented too rapidly. For a trial rate of once every 10 minutes, the vocal responses from the laboratory males usually subsided after 1 to 2 hours. However, if trials were conducted at a much slower rate, for example once per hour, then a stable calling level could be maintained within daily sessions of 10 to 14 hours for indefinite periods of time. It was on this basis that a trial rate of one per hour was adopted in testing all subsequent acoustic stimuli.

7.3 Discussion

The results of the experimental sessions involving natural mating calls have indicated a species specificity in the evoked vocal response of the male bullfrog. In later studies with natural calls (see Chapter 10), we found this response to be even more selective. Not only was it evoked only by bullfrog mating calls, but it was evoked only by certain bullfrog mating calls. Playbacks of mating calls recorded from the laboratory males have, in a number of instances, evoked a high degree of calling. However, the mating calls of certain of the laboratory males, as well as some of the field-recorded bullfrog mating calls, have reliably failed to initiate a significant amount of calling. To the human listener, most of these calls sound similar. Such selective calling by the male bullfrog indicates a fairly precise auditory discrimination. Since the females did not vocally respond in the experimental sessions, we are unable to ascertain whether the discriminatory capabilities of the male are characteristic of the female.

Chapter 8

A MODEL RELATING THE EVR TO THE ACOUSTIC STIMULUS

The experimental calling sessions with natural mating calls indicated that the evoked vocal response of the male bullfrog is highly selective. Our research has employed this vocal response as a behavioral measure of the bullfrog's ability to discriminate among different types of synthesized bullfrog mating calls. We have acquired sufficient insight into the basis of this discrimination to propose a model that allows prediction of the degree of evoked calling to a given sound. It seems more than coincidental that the model closely relates to the neural coding of acoustic stimuli, as demonstrated by electrophysiological studies of patterns of fiber activity in the eighth nerve of the bullfrog.

In a study such as this, there are a great variety of parameters that one can investigate, such as number of croaks per call, silent interval between croaks, croak duration, croak rise and fall times, croak spectral structure, and croak waveshape and waveform periodicity. A detailed study of each of these parameters was beyond the scope of our research. Consequently, we have focused our attention on croak spectral structure and waveform periodicity. This choice was taken in the belief that these are perhaps the more critical factors within the bullfrog's mating call. Furthermore, recent electrophysiological findings had revealed neural relationships involving the coding of these parameters.

The evoked-calling sessions involving synthetic mating calls will be discussed in the next chapter. A description of the model precedes their discussion in order that we might gain more insight into the interrelations of the experimental results. The model provides the framework with which to interpret the bullfrog's response to the different classes of synthetic stimuli. An evaluation of the model will be found in Chapter 10.

The spectral properties of the model are best described in terms of three frequency regions: a low-frequency region (L) centered approximately at 200 cps, a mid-frequency region (M) centered approximately at 500 cps, and a high-frequency region (H) centered approximately at 1400 cps. The model proposes that a given sound, in order to evoke the vocal response of the male bullfrog, must possess energies, with amplitudes above minimum values, in both L and H. The requirement of concomitant energy in L and H is essential. Energy in either region alone, regardless of amplitude, will not evoke calling.

The nature of the low-frequency energy is crucial. First, its spectral components must be clearly discernible; that is, the spectral envelope of the components in L should have a peak amplitude equal to, or preferably greater than, that of the components in M. Second, the optimum location for the spectral peak in L is in the neighborhood of 100 to 200 cps. The model predicts that the degree of evoked calling will monotonically decrease as the location of this spectral peak is increased above 200 cps (i.e., toward region M).

The high-frequency energy in H is not critical: all that is essentially required is that sufficient energy be present. The peak amplitude of the spectral envelope in H, relative to that in L, can vary over wide limits, from at least -40 db to +10 db, without noticeably affecting vocal performance. However, an excessive contribution in H will affect calling: if the amplitudes of the spectral components in H are increased above those in L by 30 db or more, then calling will be significantly reduced.

The model incorporates a suppressive interaction between energies in regions M and L. It predicts that the degree of calling will rapidly diminish as the amplitudes of the spectral components in M are increased above those in L. The most effective mid-frequency for this suppression is considered to be 500 cps. If its amplitude exceeds that of the spectral peak in L by 10 db, calling will be partially suppressed; if its amplitude exceeds that of the peak in L by 20 db or more, then suppression is complete. Furthermore, its effectiveness becomes more pronounced as the spectral peak in L is shifted from 200 cps toward 500 cps.

The suppression of calling by the energy in region M is not dependent on the energy in region H. The model predicts that, for those cases in which calling has been suppressed by mid-frequency energy, it cannot be regained by an independent shift in the relative amplitude of the high-frequency energy. Rather, calling can be retrieved only by an increase in the amplitude of the spectral components in L relative to those in M.

For a given spectral envelope satisfying the requirements of the model, a repetition rate of approximately 100/sec in the temporal waveform is judged to be optimum for evoking the bullfrog's vocal response. The degree of calling will be somewhat less, yet still appreciable, for waveform fundamentals above or below this optimum rate. In fact, the model proposes that a continuous spectrum, formed by filtering white noise, can evoke a significant amount of calling. Since such a sound is not quasi-periodic, the ability of an acoustic stimulus to evoke the vocal response depends far more on spectral than on temporal structure.

In summary, the more salient features of the model are the following:

1. A low-frequency region (L) centered approximately at 200 cps, a mid-frequency region (M) centered approximately at 500 cps, and a high-frequency region (H) centered approximately at 1400 cps.
2. To evoke calling, sufficient energy in both L and H must be present.
3. Spectral amplitudes in H, relative to those in L, can vary over a wide range (from -40 db to +10 db) without affecting calling performance.
4. The introduction of sufficient energy in M can suppress calling.
5. If calling is suppressed by energy in M, it can be regained only by an increase in the amplitude of energy in L relative to that in M.
6. In evoking the vocal response, the optimum repetition rate within the acoustic waveform is approximately 100/sec.

Chapter 9

THE EVR TO SYNTHETIC MATING CALLS

The evoked vocal responses of the terrarium males to natural mating calls provided the guidelines for the calling sessions with synthetic bullfrog mating calls. By a standardization in experimental technique, a quantitative measure of calling to a given sound has been obtained. This measure has permitted comparisons to be made of the relative degree of evoked calling to the different synthetic stimuli.

It is our opinion that the significance of the evoked vocal response to the synthetic mating calls parallels that to the natural mating calls. That is, the laboratory males called in response to certain synthesized sounds presumably because they accepted those sounds as the calling of another male bullfrog. The reasons for this presumption originate in the following observations. First, the evoked vocal responses to the synthetic mating calls were identical to those evoked by the natural bullfrog mating calls (i.e., only the males responded with their distinctive mating call). Second, the laboratory males called very selectively (as will be seen in the subsequent discussion of the experimental results) to certain synthesized sounds but not to others. Third, the males also called very selectively (as has already been discussed in Chapter 7) only to certain mating calls of their own species and not to others. Fourth, the physical parameters of those synthetic and natural mating calls to which the animals did respond were similar and had many properties in common. And fifth, a concentrated study of evoked calling to over ninety synthetic sounds has led to the model in the preceding chapter whereby calling to a given sound can be predicted. This model successfully explains the selectivity of the evoked vocal response to natural bullfrog mating calls, as will be discussed in Chapter 10.

9.1 Methods

The synthesis of the bullfrog's mating call has been described in Chapter 5. In our preliminary studies with these synthetic sounds, we found that the laboratory males reliably answered only certain of these stimuli. So that a comparison of calling to these sounds could be made, a standard presentation paradigm was adopted. Trials were conducted in each terrarium approximately once per hour, with random variations of up to ±20 minutes superimposed to reduce hopefully any conditioning aspects of trial rate. Experimental sessions were conducted daily, seven days a week, for an average in each terrarium of twelve trials per day.

As in the previous studies with natural calls, each trial involved a prestimulus five-minute monitor of background noise and spontaneous calling. If the noise level had been stable and no calling had occurred, then a given synthetic stimulus was initiated in a temporal sequence of several calls, each comprised of a number of identically synthesized bullfrog mating croaks. The sequence was: 3-croak call, silent interval of 12 seconds; 7-croak call, silent interval of 9 seconds; 5-croak call, silent interval of 10.5 seconds; 6-croak call, silent interval of 15 seconds; 8-croak call, silent interval of 7.5 seconds; and a 4-croak call. All croaks within each call were separated by 0.7 second; each croak lasted for 0.8 second and had symmetrical rise and fall times of 250 milliseconds. The entire presentation required a total of 103 seconds.

If, during the stimulus presentation, any male issued a mating call (the criterion being a minimum of three repetitive croaks), then the presentation was immediately terminated. On the other hand, if no calling occurred, the presentation progressed to completion. The choice to terminate upon calling was taken in an effort to reduce the possibility of habituation to these sounds. A positive score was assigned to a trial if at least one mating call, of three or more croaks, was evoked during either the stimulus presentation or the following poststimulus interval of 80 seconds; a negative score was assigned if calling did not occur during both of these periods. A trial was terminated upon the completion of the poststimulus interval, and another was not presented until approximately one hour's lapse.

The different synthetic stimuli were generally tested in groups of 5 to 10. Within each group, the stimuli were presented in random order during consecutive experimental sessions until each stimulus had been presented for a minimum of 10 to 20 trials. An evoked-calling level, based on the percentage of positive trials, was then assigned to each stimulus within that group. In this manner, a quantitative measure of evoked calling to each synthetic mating call was obtained which enabled a comparison to be made between the different stimuli.

Experimental controls must be incorporated into every behavioral study in order to insure that the observed sensory performance is in fact due to the presented stimulus and not to extraneous experimental conditions. An effort has been made to maintain constancy of laboratory conditions throughout this study. Visitors, other than the experimenter, were seldom allowed entrance into the laboratory. Extraneous disturbances were held to a minimum (to the extent that this was possible in a wooden frame structure housing an interdepartmental laboratory). The terrarium animals were rarely handled or physically disturbed. Watering and feeding routines were always conducted in the same manner. Laboratory lighting levels were identically cycled daily, and terrarium temperatures were held relatively constant (67 to 72°F).

All experimental sessions were conducted as identically as possible. Trials generally began each day at 8:30 AM and continued hourly until 7:30 PM in each terrarium. A trial was never conducted in the presence of transient noise or otherwise unwanted interference. If such a disturbance happened to occur in the middle of a stimulus presentation, the results of that trial were discounted. The order of presentation of different stimuli was independently randomized in each terrarium.

Precautions were taken to insure that stimulus presentations were not masked by background noise. For most stimuli, this was achieved with an acoustic level of approximately 64 db SPL (re 0.0002 dyne/cm^2). A slightly higher acoustic level was needed for a few other stimuli. (See Chapter 6 for the criterion of choosing a suitable acoustic level.) Representative stimuli of the different classes of synthetic mating calls were further presented at several different levels to insure that calling performance was not dependent on the acoustic level. The sound-pressure levels employed for each stimulus will be included in the discussions of the experimental results.

Two different terrariums, each stocked with animals of different geographical origin and housed in separate laboratory rooms, permitted cross-comparisons of evoked-calling results. The performance of the two colonies has been very similar. Calling levels in the two terrariums, for a sufficient number of trials, generally differed by no more that ±10% for any given stimulus. In no instance was it ever found that calling to any of the ninety-odd stimuli reliably occurred in one terrarium but not in the other. As a consequence, trial results for the two colonies have been lumped together to give a composite calling level for each synthetic mating call.

The result of these efforts has been a remarkable constancy in calling performance. Tests of certain representative stimuli, when conducted at disjoint intervals during the course of our experimentation, repeatedly yielded very similar calling levels. Based on a minimum of 10 to 20 trials for each disjoint repetition, levels seldom varied by more than ±10% and never by more than ±20%. Since the calling levels were so stable, all trials for these stimuli have been combined to give an average calling level.

9.2 Results

9.2.1 Necessity of Energy in Regions L and H. In order to evoke the vocal response from the terrarium males, sufficient

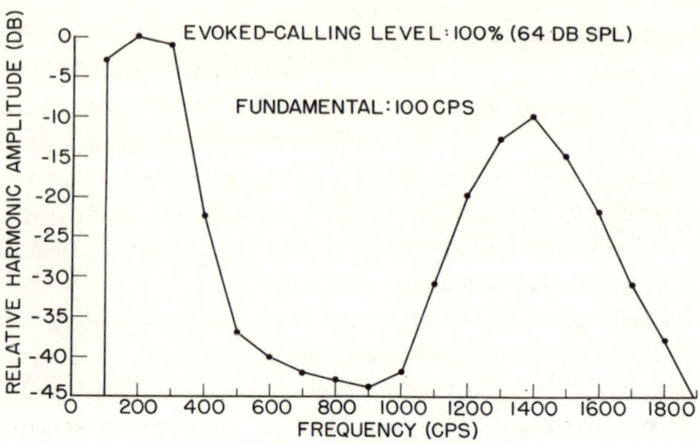

Figure 9.1. Period-synchronous spectrum of Synthesized Bullfrog Mating Croak Number 1 (SBFMC-1). The spectrum shows the relative amplitudes of the harmonics in the periodic waveform of this synthesized croak. (See text for details of synthesis.) Measurements were taken at the input to the terrarium loudspeakers with a Hewlett Packard 302A wave analyzer (electronic switch disabled). The spectrum therefore corresponds to a spectrographic section integrated over a single period of the periodic waveform (i.e., a "period-synchronous" spectrum).

energy had to be simultaneously present in both a low-frequency region (L), centered approximately at 200 cps, and a high-frequency region (H), centered approximately at 1400 cps. A harmonic line spectrum, whose relative amplitudes constitute a spectrographic section through the synthesized croaks of an acceptable synthetic mating call,* is shown in Figure 9.1. The bimodal envelope of the harmonic components emphasizes the concentration of energy in the two disjoint regions: a spectral peak in L occurs at 200 cps, and a peak in H occurs at 1400 cps. This stimulus, as well as most of the other line spectra, was generated by the parallel addition of the filtered outputs of two of the POVO-type RLC resonant circuits (Chapter 5). The repetition rate of the pulse-train excitation of the resonators was set at 100/sec. Therefore, the temporal fine structure within the synthesized croak waveform had a periodicity of 100/sec (see Figure 5.8 for

*The relative amplitudes of the spectral components in the electrical waveform input to the terrarium loudspeakers were measured for each synthetic stimulus and were verified by a similar analysis of the resultant acoustic waveforms. These measurements were made with a Hewlett Packard 302A wave analyzer and a Kay Sona-Graph.

some representative waveforms), and harmonics in the line spectrum occurred every 100 cps.

The relative amplitudes of the harmonics in the Synthesized Bullfrog Mating Croak Number 1 (hereafter abbreviated SBFMC-1) of Figure 9.1 show the peak in L to exceed the peak in H by 10 db. When this synthetic croak was presented in the standardized sequence of several mating calls at an acoustic level of approximately 64 db SPL, calling was evoked from the laboratory males during every trial (i.e., an evoked-calling level of 100%). This calling score was computed on the basis of 10 separate trials in each terrarium for a total of 20 trials. Such performance was comparable to the greatest calling observed to natural bullfrog mating calls. The presentation intensity had little effect on calling performance: a score of 100% was obtained for all acoustic levels from 58 to 82 db SPL. Presentations above 82 db SPL

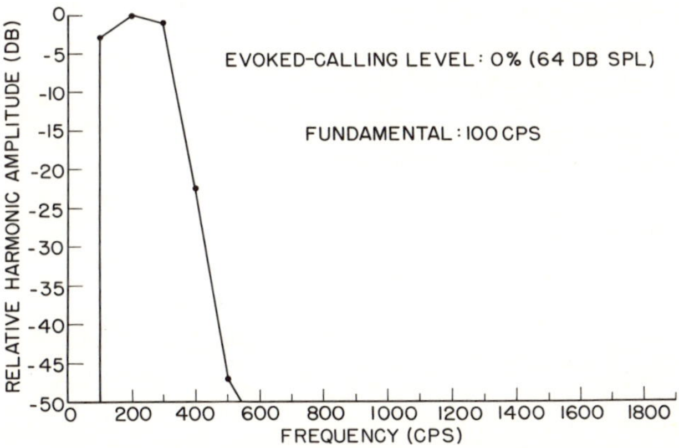

Figure 9.2. Period-synchronous spectrum of SBFMC-1A.

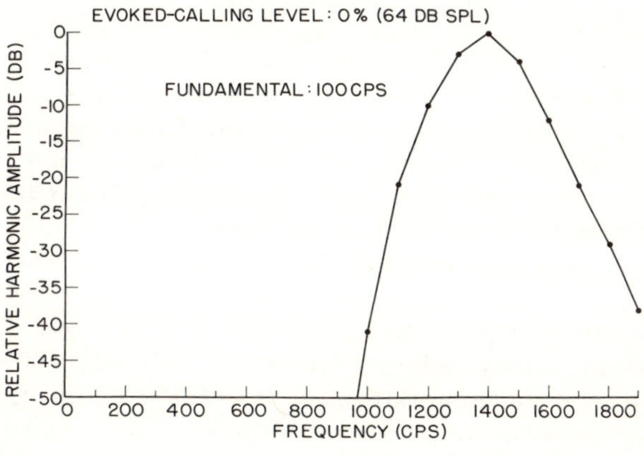

Figure 9.3. Period-synchronous spectrum of SBFMC-1B.

were not attempted for fear of frightening the animals; presentations below 58 db SPL were not investigated because of possible masking interference by background noise.

Trials involving the synthesized croaks of Figures 9.2 and 9.3 clearly demonstrated that energy in both L and H was necessary to initiate calling. The stimulus of Figure 9.3 (SBFMC-1A) was derived from that of Figure 9.1 (SBFMC-1) by low-pass filtering at 400 cps (102 db/oct) and represents the low-frequency components of that sound. There was no calling whatsoever to SBFMC-1A at presentations of 64 db SPL as well as at other levels of 58 to 82 db SPL. Figure 9.3 (SBFMC-1B) represents the high-frequency components of Figure 9.2 and was obtained by high-pass filtering at 1000 cps (102 db/oct). To this high-frequency energy alone, there was absolutely no calling for acoustic levels of 58 to 82 db SPL. The results for these stimuli are striking. Vocal responses could not be evoked by the low-frequency harmonics alone or by the high-frequency harmonics alone. Yet, when the components of these two regions were superimposed, a maximum calling level of 100% was attained.

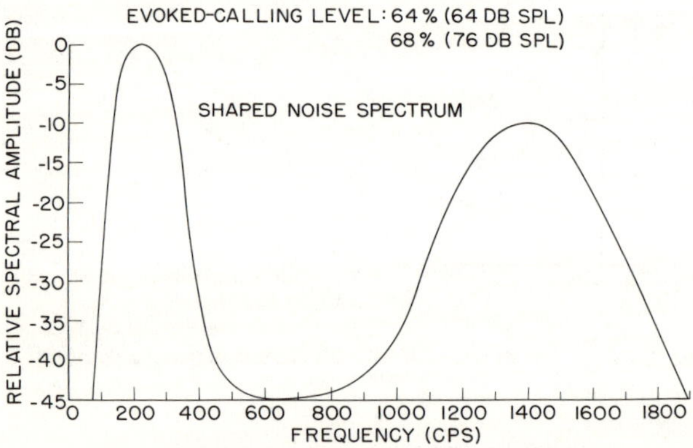

Figure 9.4. Continuous spectrum of SBFMC-2. This croak was synthesized by exciting two of the synthesis resonators with white noise from the noise generator. The envelope of its spectrum corresponds to the frequency response of the synthesis system (from the output of the noise generator to the input of the loudspeaker, electronic switch disabled).

The requirement of concomitant energy in regions L and H has been verified by using other classes of stimuli. Figure 9.4 (SBFMC-2) shows the spectral envelope for a synthetic croak comprised of a continuous spectrum. This stimulus was formed by exciting two of the synthesis resonators with white noise rather than periodic pulse trains. Its spectrum was additionally shaped by band-pass filtering of each resonator output prior to summation. The spectral envelope for this stimulus is very similar to that for the harmonic line spectrum of Figure 9.1 (SBFMC-1). Again the relative amplitude of the spectral peak at 200 cps in L exceeded that of the peak at 1400 cps in H by 10 db. Presentation of this "noise-croak" evoked vocal responses from the laboratory males. Based on 25 trials at 64 db SPL, a calling level of 64% was scored; for 25 trials at 76 db SPL, a level of 68% resulted. In fact, for a minimum of 10 to 20 trials, an evoked-calling level of 60 to 75% corresponded to all presentations between 58 and 82 db SPL. These noise stimuli lacked a repetitive temporal fine structure. Although they did not evoke as high a calling level as the harmonic line spectrum with a fundamental of 100 cps, there was nevertheless a significant response to their presentation. Such performance suggests an emphasis on the spectral distribution of stimulus energy, as contrasted with the temporal fine structure, in the bullfrog's detection of his mating call.

When the continuous spectrum of Figure 9.4 was low-pass filtered at 400 cps or high-pass filtered at 1000 cps (102 db/oct), calling failed to occur for all trial presentations of 58 to 82 db SPL. It was only when the contributions of regions L and H were combined that vocal responses could be evoked.

Further evidence that both the low-frequency region and the high-frequency region had to be excited in order to initiate calling was offered by stimuli comprised of tonal combinations. Neither a sinusoid of 250 cps alone nor of 1300 cps alone was capable of producing calling (58 to 82 db SPL). Yet in combination (amplitude of the 250-cps tone +10 db re amplitude of the 1300-cps tone), the composite stimulus evoked a calling level of 67% (15 trials, 64 db SPL). Furthermore, though the low-region harmonics of Figure 9.2 (SBFMC-1A) or the high-region harmonics of Figure 9.3 (SBFMC-1B) by themselves evoked no response, the addition of a single appropriate tone to either did evoke a response. If a high-frequency line was introduced to the components of Figure 9.2, calling could then be initiated. For example, the addition of a 1300-cps tone, in relative amplitude of 10 db below the 200-cps peak in L, resulted in a calling level of 90% (10 trials, 64 db SPL). Similarly, the sole addition of a 250-cps tone to the components of Figure 9.4 (amplitude of 250-cps tone +10 db re amplitude of 1400-cps peak in H) changed the calling level from 0% to 70% (10 trials, 64 db SPL).

Our experimental results involving the various classes of synthesized mating croaks have all supported the conclusion that energy must be present in both regions L and H in order to evoke the vocal response of the male bullfrog. In no instance have we been able to evoke calling from our laboratory animals by the presentation of energy, regardless of amplitude, in either region alone.

9.2.2 Signal Requirements in Region H. A family of synthetic croaks was generated in which the spectral envelopes over the low-frequency region and over the high-frequency region were held fixed in shape but allowed to vary in relative amplitude. Spectral peaks in L and H were again located at 200 cps and 1400 cps. These stimuli were realized by summing different voltage ratios of resonator outputs. For this entire sequence, a pulse-train excitation of 100 pulses/sec was used. Therefore, the line spectra possessed a harmonic relationship of 100 cps, and the temporal waveforms exhibited a periodicity of this same rate.

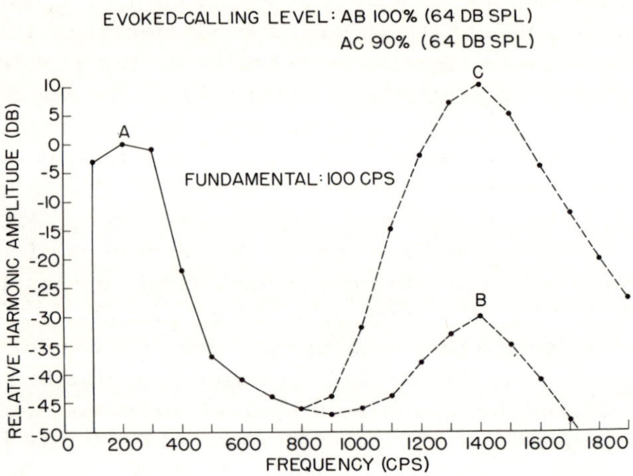

Figure 9.5. Period-synchronous spectra of SBFMC-3AB and SBFMC-3AC.

Figure 9.5 gives the relative harmonic amplitudes and calling levels for two such synthesized croaks. In SBFMC-3AB the spectral peak in H was down by 30 db relative to the peak in L, whereas in SBFMC-3AC the peak in H exceeded that in L by 10 db. For trial presentations of 64 db SPL, the calling levels were comparable. In fact, for this entire family of synthetic croaks, the spectral peak in region H, relative to the peak in region L, could vary in amplitude from -40 to +10 db without substantially reducing the evoked-calling level. It was only when the 1400-cps peak was raised by 30 db or more above the 200-cps peak that calling could be severely affected. These results are summarized in Table 9.1 below. Acoustic levels for all trials were adjusted to 64 db SPL.

The last entry in the table requires further explanation. For this stimulus, in which the 1400-cps peak in H exceeded the 200-cps peak in L by 30 db, calling seldom occurred to presentations at 64 db SPL. This was also the case for higher acoustic levels

Table 9.1. Calling Levels for Various Amplitudes of the Spectral Peak in H (1400 cps) Relative to that in L (200 cps)

Relative Amplitude of Spectral Peak in L (db)	Relative Amplitude of Spectral Peak in H (db)	Evoked-Calling Level (%)
0	-40	95
0	-30	100
0	-10	100
0	0	90
0	+10	90
0	+30	10

as well (up to 82 db SPL). Therefore, the decrease in calling should not be attributed to subthreshold stimulation in region L nor to masking of the low-frequency energy by background noise. Rather, it appears that simply the large imbalance of energy in region H, relative to that in region L, caused the decrease in calling.

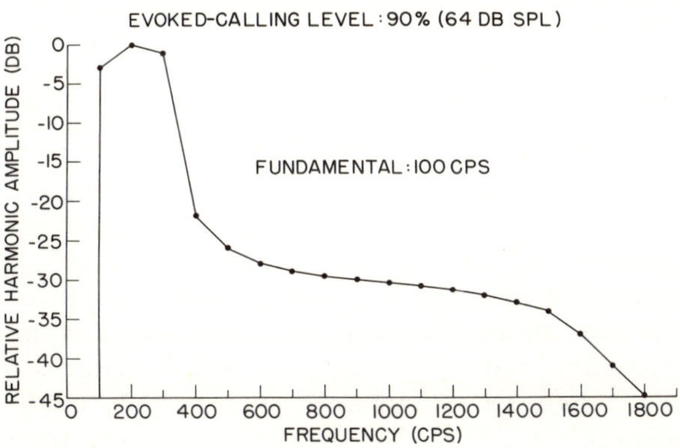

Figure 9.6. Period-synchronous spectrum of SBFMC-4.

In evoking the male's vocal response, the energy distribution in region H need not necessarily exhibit a relative amplitude peak. That is, the spectral envelope for an acceptable synthetic croak can be unimodal, with a single peak in L, as shown by the line spectrum in Figure 9.6 (SBFMC-4). For this synthesized mating croak, a fundamental periodicity of 100/sec was again chosen. Note that the spectral envelope does not show a relative peak in region H, and a corresponding center frequency in this region cannot be identified. Nevertheless, this stimulus evoked a high degree of calling (90%, 64 db SPL).

Trials involving continuous spectra have verified that the relative amplitude of the energy in region H can vary over wide limits and that spectral peaking in this region is not essential. The noise-croak of Figure 9.7 (SBFMC-5) represents the extreme case in which the contribution in H does not show a relative peak: the spectral envelope is similar to that for the harmonic line spectrum in Figure 9.7. A calling level of 67% was evoked by its acoustic presentation at 76 db SPL. (This higher acoustic level was chosen so as to reduce any possibility of high-frequency masking by the background noise.) This calling performance is almost identical to that for the continuous spectrum of Figure 9.4 (SBFMC-2) in which a spectral peak in H is clearly evident.

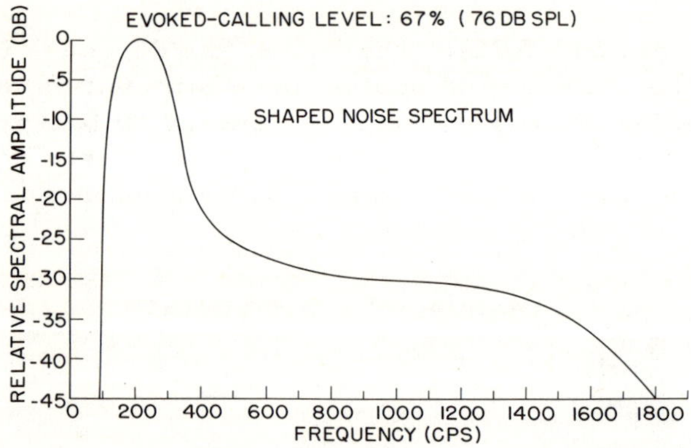

Figure 9.7. Continuous spectrum of SBFMC-5.

We have concluded from these calling sessions with harmonic line and continuous spectra that the signal requirements in region H are not critical in evoking the male's vocal response. All that is essentially required of this region is the presence of sufficient energy. The converse situation does not apply to the energy in the low-frequency region. Vocal responses are highly dependent on the existence of a spectral peak in L. These results are discussed next.

9.2.3 Signal Requirements in Region L. In comparing those natural bullfrog mating calls to which calling could be evoked, we found that their spectra generally had a distinct peak in the neighborhood of 200 to 400 cps. To test the significance of this finding, we therefore generated a sequence of synthetic mating croaks in which a single peak in their spectra was systematically varied from 100 to 1000 cps. Figure 9.8 (SBFMC-6) shows the spectral envelope, with a peak at 300 cps, for one such stimulus. Synthesis was realized by a superposition of the filtered outputs, in appropriate amplitude ratios, of three resonators to 100 pulses/sec excitation. (The band-pass filtered output of one of the resonators was dominant in order to produce a sharp spectral peak.) The spectra for all of the stimuli were similar to that of

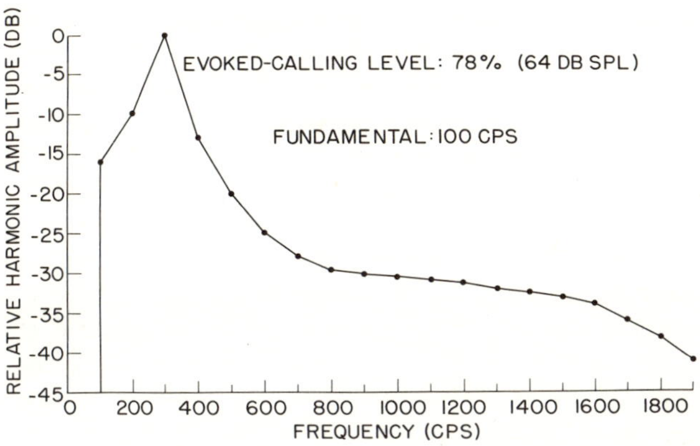

Figure 9.8. Period-synchronous spectrum of SBFMC-6. Note that the envelope of the harmonic spectrum has a low-frequency peak at 300 cps. In each synthesized croak of the corresponding sequence, this peak occurred at a different harmonic of the fundamental.

Figure 9.8, except that the low-frequency peak occurred in each at a different harmonic of the fundamental.

Evoked-calling levels, based on 10 to 20 separate trials for each stimulus, were obtained for this entire sequence of synthetic mating croaks. Acoustic presentations were maintained at 64 db SPL in all trials. The calling percentages corresponding to the different locations of the low-frequency peak are given in Figure 9.9. These data show that the evoked vocal response is highly dependent on a spectral peak in L. That is, the calling level was greatest for the low-frequency peak at 100 or 200 cps; it decreased rapidly and monotonically while this peak approached the mid-frequency region (M) centered about 500 cps. In inter-

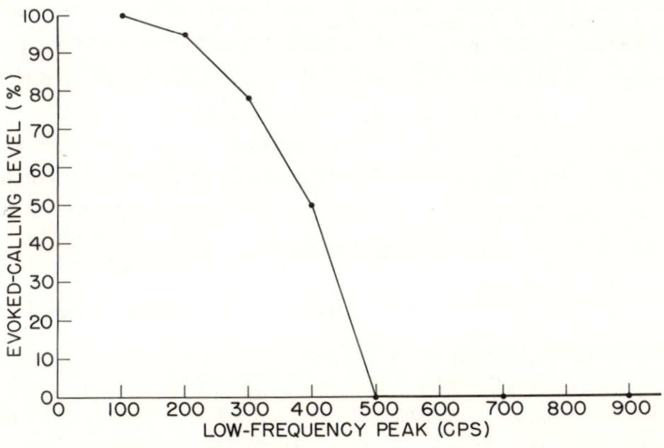

Figure 9.9. Evoked-calling levels for different locations of the low-frequency peak (Figure 9.8).

preting these results, we should recognize (from a consideration of Figure 9.8) that, as the location of the peak was increased above 200 cps, stimulation of region M became progressively greater while stimulation of region L became progressively less. This transition is very significant. We shall find in our subsequent discussions that a spectral component in region M can suppress calling to an extent dependent on its amplitude relative to those in region L. As a result, the degree of calling to a given stimulus will be optimum if a relative peak in its spectrum occurs in L. A monotonic decrease in calling level should therefore be expected of a shift from L toward M in the location of this peak.

Calling levels for continuous spectra, having these same spectral envelopes, yielded similar results. For these noise-croaks, the degree of calling was greatest for the low-frequency peak at 150 to 200 cps (62 to 67%, 76 db SPL). Again, calling monotonically decreased as the location of the peak was increased above 200 cps; vocal responses could not be evoked if the peak occurred in the neighborhood of 500 cps (70 to 76 db SPL).

Synthetic croaks, composed of two tones, were also tested. To a 1300-cps sinusoid, various low-frequency tones were added (amplitude of low-frequency tone +10 db re amplitude of 1300-cps sinusoid). The addition of a 200-cps tone evoked the most calling (67%), a 300-cps tone evoked considerably less calling (36%), and a tone of 400 cps, or above, evoked zero calling. All tonal combinations were presented at 64 db SPL.

These experimental sessions have demonstrated that the degree of evoked calling to a given stimulus is predominantly dependent upon the spectral peak of its low-frequency energy. The male bullfrog must therefore be highly tuned to frequencies between 100 and 500 cps. This sensitivity accounts, in part, for his very selective vocal response to certain sounds that otherwise have quite similar characteristics.

9.2.4 Suppression by Energy in Region M. In testing various synthetic mating croaks, we noticed that the introduction of sufficient energy between 400 and 700 cps would suppress the vocal responses from the males of our laboratory colonies. This suppression suggested a close relationship, as will be discussed in Chapter 10, to Frishkopf and Goldstein's[63] electrophysiological finding of inhibition in the bullfrog's eighth nerve response to low-frequency tones. Since frequencies in the neighborhood of 500 cps had been found to be the most effective for inhibiting the neural responses, we focused our attention on this frequency and its effect on the evoked vocal response.

A sequence of harmonic line spectra was synthesized in which the location of a low-frequency peak again varied systematically, as in the preceding study, but with the notable difference of the addition of a second peak at 500 cps. In this entire sequence, the amplitude of the mid-region peak at 500 cps was maintained equal

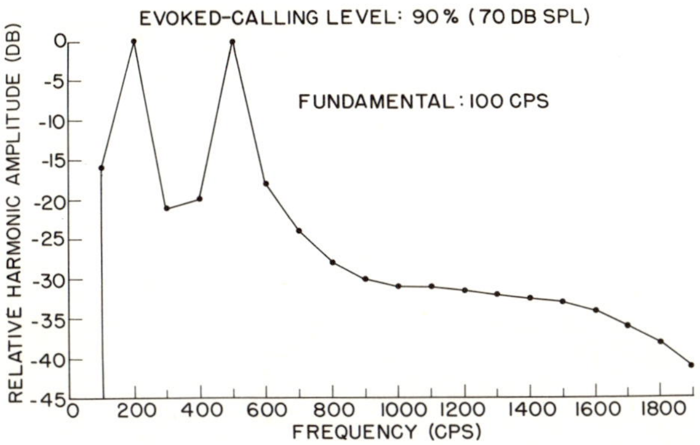

Figure 9.10. Period-synchronous spectrum of SBFMC-7. In the corresponding sequence of synthesized croaks, the peak at 500 cps was fixed. This spectrum shows the case for the equal-amplitude, low-frequency peak at 200 cps.

to that of the low-frequency peak. The distribution of harmonics in the high-frequency region was again relatively uniform and similar in all of these stimuli. Figure 9.10 (SBFMC-7) shows the spectrum, for the low-frequency peak at 200 cps, of one such stimulus. Each of the two spectral peaks was realized by sharply band-pass filtering a separate resonator output; the filtered outputs of two additional resonators were required to produce the relatively flat envelope over the high-frequency region. The four resonators were excited simultaneously by a periodic pulse train of 100 pulses/sec. The spectral envelopes for all stimuli were identical to that of Figure 9.10, except that the low-frequency peak in each occurred at a different harmonic of the fundamental.

Evoked-calling levels were determined on the basis of 10 to 20 trials, conducted at 70 db SPL, for each stimulus. The results are shown in Figure 9.11. (Also included, for comparison, are the calling levels for the same low-frequency peaks but with the 500-cps peak absent, viz., Figure 9.9.) For the low-frequency peak located at 100 or 200 cps, the introduction of the equal-amplitude peak at 500 cps had little effect on the evoked responses. As would be expected, calling again monotonically decreased as the low-frequency peak was shifted from region L toward region M. However, in this shift, the presence of the 500-cps peak significantly produced an additional suppression of calling (relative to that for the 500 cps absent). The degree of suppression could not be appreciably altered by a spectral peaking or an increase in the relative amplitude of the high-frequency energy in region H nor by an increase in the acoustic presentation level. These results suggested that the suppression of calling was due to an interaction between energies in regions L and M and was

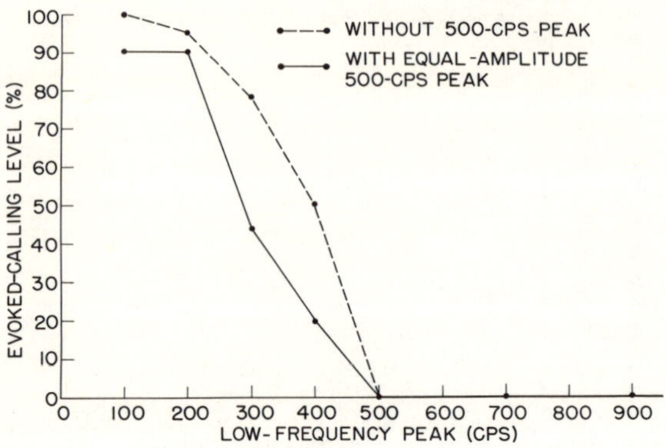

Figure 9.11. Evoked-calling levels for different locations of the low-frequency peak with and without the equal-amplitude peak at 500 cps (Figures 9.8 and 9.10).

relatively unaffected by energy in region H. We investigated the nature of this interaction in more detail by the following set of experiments.

An extensive family of croaks was synthesized in which various amplitude ratios of low-frequency to high-frequency energy were realized and to which were added different levels of a 500-cps spectral line. The energy distributions were chosen according to the spectral envelope that seemed to evoke most calling; namely, a low-frequency peak at 200 cps in L and a high-frequency peak broadly centered about 1400 cps in H. The spectra were

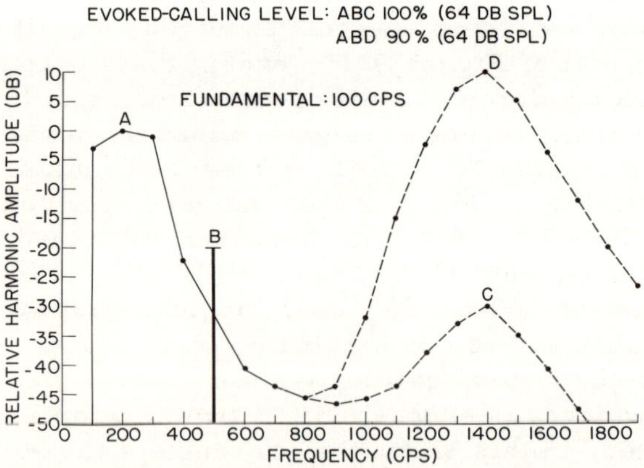

Figure 9.12. Period-synchronous spectra of SBFMC-8ABC and SBFMC-8ABD. These synthesized croaks were realized by adding a single 500-cps line to the spectra of SBFMC-3 (Figure 9.5). The frequency of this line was maintained in harmonic relationship to the fundamental.

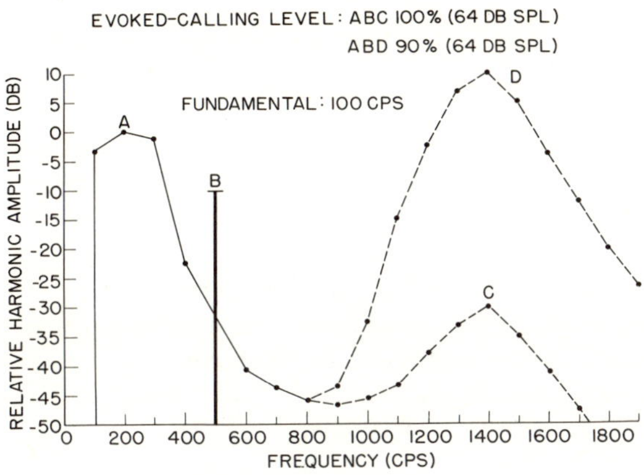

Figure 9.13. Period-synchronous spectra of SBFMC-9ABC and SBFMC-9ABD.

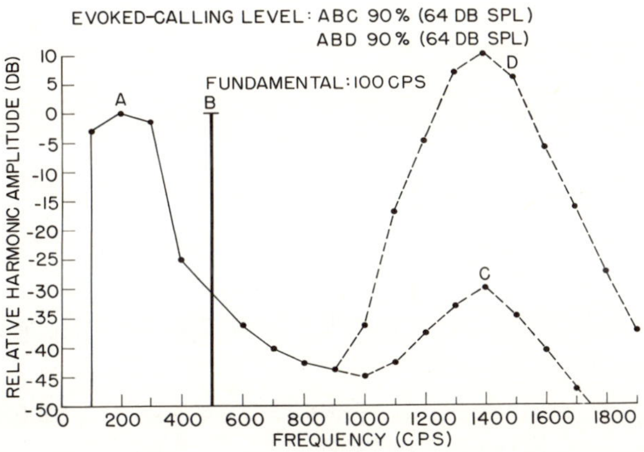

Figure 9.14. Period-synchronous spectra of SBFMC-10ABC and SBFMC-10ABD.

again harmonic with a fundamental of 100 cps. All calling scores were based on a minimum of 10 trials, and acoustic levels were held constant, unless otherwise noted, at 64 db SPL.

The introduction of the 500-cps line, of amplitude equal to or less than that of the 200-cps peak, had no measurable effect on calling. Figures 9.12, 9.13, and 9.14 indicate respective calling scores for the 500-cps component at -20 db, -10 db, and 0 db relative to the amplitude of the 200-cps peak. A comparison with the synthetic croak of Figure 9.5 (SBFMC-3), in which the 500-cps line was absent, shows essentially no difference in calling response: high calling levels of 90 to 100% apply in all cases. Furthermore, a variation of 40 db in the relative amplitude of the 1400-cps peak, which again reflects the relatively independent nature of regions L and H, had little effect on the vocal responses.

An additional increase in the amplitude of the 500-cps line produced a remarkable change. When this component was raised but 10 db above the 200-cps peak, as shown in Figure 9.15, calling declined from the previous level of 90 to 100% to a level of 20 to 30%. (SBFMC-11ABC was presented at 70 db SPL, rather than at 64 db SPL, in order to reduce any possiblity of masking by background noise.) The 500-cps component appears to exert its influence entirely on the lower-frequency energy in region L. A 40-db variation in the relative amplitude of the 1400-cps peak in H, demonstrated by SBFMC-11ABC and ABD in Figure 9.15, could not regain a high degree of calling.

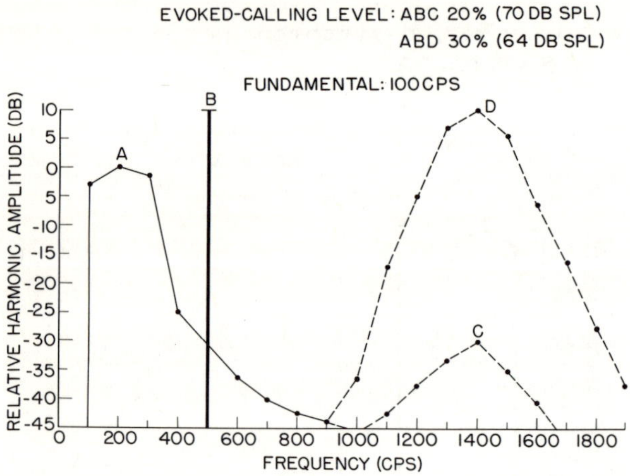

Figure 9.15. Period-synchronous spectra of SBFMC-11ABC and SBFMC-11ABD.

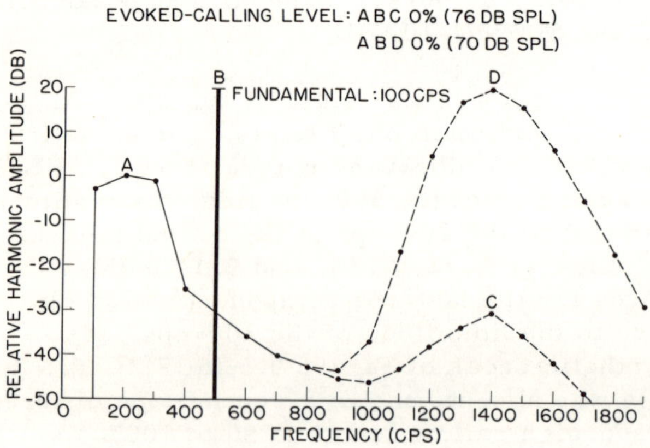

Figure 9.16. Period-synchronous spectra of SBFMC-12ABC and SBFMC-12ABD.

A still further increase in the relative amplitude of the 500-cps line, to 20 db above the 200-cps peak, caused a total suppression of calling. This result is shown by SBFMC-12ABC in Figure 9.16 (presented at 76 db SPL). A shift of 50 db in the relative amplitude of the 1400-cps peak failed to offset the suppressive effect of the 500-cps component (SBFMC-12ABC, 70 db SPL).

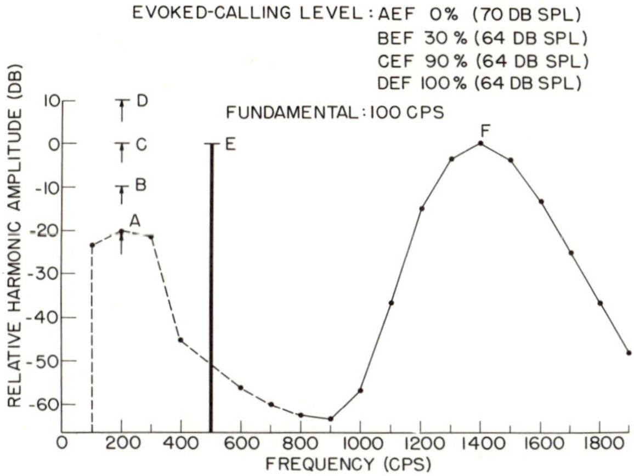

Figure 9.17. Period-synchronous spectra of SBFMC-13.

In fact, for this stimulus condition, in which the amplitude of the 500-cps line exceeded that of the 200-cps peak by 20 db, it was impossible to evoke the vocal response from the males in either terrarium colony by any change in the relative amplitude of the 1400-cps peak (for all acoustic presentation levels between 64 and 82 db SPL). Calling could only be regained by an increase in the relative amplitude of the 200-cps peak. Figure 9.17 gives the results for such a change. As the amplitude of the low-frequency peak was progressively raised in 10-db steps, from -20 to +10 db relative to the amplitude of the 500-cps line, shown respectively by SBFMC-13AEF, BEF, CEF, and DEF, calling increased from 0% to a maximum of 100%. Note especially the large change that occurred in the evoked-calling level, corresponding to the transition from BEF to CEF, as the amplitude of the low-frequency peak was made equal to that of the 500-cps line. This change was insensitive to a wide variation, from at least -40 to +10 db, in the relative amplitude of the 1400-cps peak.

A compendium of the evoked-calling results, for various relative amplitudes of the 200-cps peak, 500-cps line, and 1400-cps peak, is given in Table 9.2. Amplitudes are specified relative to that of the 200-cps peak. The acoustic presentation level, corresponding to the calling percentage shown, is also included.

Table 9.2. Effect of the 500-cps Spectral Line on the Evoked-Calling Level

Relative Amplitude of 200-cps Peak (db)	Relative Amplitude of 500-cps Line (db)	Relative Amplitude of 1400-cps Peak (db)	Evoked-Calling Level (%)	Acoustic Presentation Level (db SPL)
0	-20	-30	100	64
0	-20	0	90	64
0	-20	+10	90	64
0	-10	-30	100	64
0	-10	-10	100	64
0	-10	0	90	64
0	-10	+10	90	64
0	0	-30	90	64
0	0	-10	100	64
0	0	0	90	64
0	0	+10	90	64
0	+10	-30	20	70
0	+10	-10	20	64
0	+10	+10	30	64
0	+10	+30	0	76
0	+20	-30	0	76
0	+20	-10	0	70
0	+20	+10	0	70
0	+20	+20	0	70

Experimental sessions with tonal stimuli have offered additional evidence that the presence of 500-cps energy has a suppressive effect on the evoked vocal response. As was mentioned earlier, the combination of a 250-cps tone and a 1300-cps tone (amplitude of 250-cps tone +10 db re amplitude of 1300-cps tone) evoked a calling level of 67% (64 db SPL). The addition of a 500-cps tone to this combination could abolish calling. Figure 9.18 shows the results for such an addition. For the 500-cps line equal in amplitude to the 250-cps line, the calling level decreased from 67% to 46% (SBFMC-14ABD, 64 db SPL). A further increase of 10 db in the relative amplitude of the 500-cps line completely abolished calling (SBFMC-14ACD, 64 db SPL).

And finally, we have found that the evoked vocal responses to stimuli having a continuous spectrum can be abolished by the introduction of energy in the mid-region M. The addition of a sharply filtered (102 db/oct) narrow band of noise, extending from 500 to 600 cps, to the noise-croak of Figure 9.7 (SBFMC-5) changed the calling level from 67% to 0% (amplitude of noise band +10 db

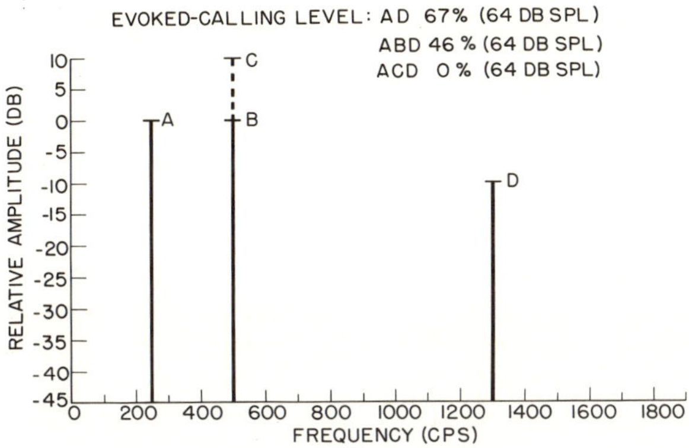

Figure 9.18. Line spectra of SBFMC-14. These croaks were synthesized by a combination of three tones: 250 cps, 500 cps, and 1300 cps. SBFMC-14AD corresponds to the case in which the 500-cps component is absent.

re amplitude of spectral peak at 200 cps, 76 db SPL).

In summary, the experimental sessions involving harmonic and nonharmonic line spectra, as well as continuous spectra, demonstrated that the presence of energy in the neighborhood of 500 cps (i.e., in region M) could suppress the evoked vocal response of the bullfrog. The degree of suppression depended upon the amplitudes of the spectral components in region M relative to the amplitudes of the components in region L. Calling was essentially absent for all stimulus conditions in which the amplitude of the 500-cps component exceeded that of the spectral peak in L by 10 db or more. This suppression appears to be caused by an interaction between the energies of only regions L and M: wide variations (40 to 50 db) in the relative amplitude of the spectral peak in region H had little effect. For those cases in which calling was suppressed by energy in region M, it could be regained only by increasing the relative amplitudes of the spectral components in region L.

9.2.5 Waveform Periodicity. We have restricted our investigations of the temporal fine structure of the acoustic stimulus to that of waveform periodicity. Although this choice was dictated by the unavoidable phase distortion in the terrarium acoustics, we do not consider it to be a severe restriction. Our analysis of several natural bullfrog mating calls had revealed that, whereas the waveshapes in different croak waveforms varied widely, the temporal fine structure in all of these calls had a similar fundamental frequency (Chapter 5). Furthermore, in several preliminary sessions involving natural and synthetic mating calls, we found that we could not appreciably alter the evoked-calling levels for these stimuli by reversing their playback on a tape recorder. A backward presentation preserved the temporal periodicity and spectral distribution (except for a sign reversal of phase angle) but, since these signals were not symmetrical, the waveshape in the acoustical waveform was obviously quite different. (E.g., compare the waveshapes in Figures 5.1 to 5.3 and 5.8 with their mirror images.) These results suggested that the periodicity in the croak waveform might play a more important role in evoking the bullfrog's vocal response.

As mentioned in Chapter 5, the bullfrog's mating croak has a rather stereotyped quasi-periodicity of approximately 100/sec. It was on this basis that a fundamental of 100 cps was chosen for all of the harmonic line spectra previously described. Synthesized mating croaks having other periodicities were briefly investigated in our experimental sessions. For these stimuli, the spectral envelopes were held constant while the repetition rate of the pulse-train input to the resonators was varied from 25 to 200/sec. Continuous spectra, generated by exciting the resonators with white noise, were included in these experimental sessions.

This study revealed that the evoked-calling level would be greatest, for a given spectral envelope, if the corresponding temporal waveform had a periodicity of approximately 100/sec. The rank order, in decreasing ability to evoke the vocal response, for the different fundamentals and noise was (1) 100 cps, (2) 150 cps, (3) 200 cps, (4) noise, and (5) 50 cps or lower. For example, for the representative spectral envelope of Figure 9.4, evoked-calling levels were obtained on the basis of 25 trials for each mode of excitation. All trials were conducted at 64 db SPL. The calling percentages, corresponding to the different fundamentals, for these synthetic mating croaks were (1) 100 cps — 100%, (2) 150 cps — 92%, (3) 200 cps — 76%, (4) noise — 64%, (5) 50 cps — 48%, and (6) 25 cps — 0%. To the stimulus with a fundamental of 25 cps, we found that calling could not be evoked for all acoustic levels of 58 to 82 db SPL.

Synthesized mating croaks, having the spectral envelope of Figure 9.7, yielded similar results. For approximately 20 trials with each stimulus, the evoked-calling levels were (1) 100 cps — 90%, (2) 150 cps — 90%, (3) 200 cps — 80%, (4) noise — 67%, (5) 50 cps — 55%, and (6) 25 cps — 0%. All trials, except those involving the noise-croak (76 db SPL in order to minimize masking) were conducted at 64 db SPL. Again, calling could not be initiated by increasing the acoustic level of the stimulus with fundamental of 25 cps.

Our finding of greatest calling to those stimuli with fundamental frequencies of 100 cps is not surprising. This is approximately the same periodicity as that found in the natural croak waveform of the bullfrog's mating call.

That calling diminished so markedly, for waveform periodicities of 50/sec and below, might be related to certain bullfrog territorial calls (Appendix A). The waveforms of these calls often show a periodicity between 15 and 30/sec. It may be that a given sound, as the repetition rate of its temporal fine structure is decreased, takes on a different biological significance — a significance in which a vocal response is not normally given. At the same time, it must be recognized that our synthetic mating calls were generated by exciting resonant circuits with periodic pulse trains. As such, for lower repetition rates, the temporal waveform becomes more pulsatile. By an appropriate adjustment of the phases of the harmonics, it could be that a less pulsatile sound, having a low fundamental frequency, could be produced that would evoke the bullfrog's vocal response. We did not investigate this possibility.

Synthetic mating croaks, synthesized by filtering white noise, have reliably evoked a calling level of 60 to 75%. Since the temporal fine structure within these sounds is random, their ability to evoke the male's vocal response indicates an emphasis on the spectral distribution of the acoustic stimulus. Synthesized mating croaks, having a nonharmonic line spectrum, have also shown this to be the case. A combination of five equal-amplitude sinusoids, with frequencies of 105, 195, 305, 1345, and 1450 cps, evoked a calling level of 75% (12 trials) when presented at 70 db SPL. The temporal waveform for this stimulus had no repetitive fine structure. When a tone of 500 cps, with a relative amplitude of +10 db, was added to this line spectrum, the evoked-calling level dropped to zero (64 to 82 db SPL). It would appear that, as long as the temporal periodicity is above a certain rate (approximately 50/sec) or else completely absent, the spectral characteristics of a given sound predominate in evoking the vocal response of the male bullfrog.

PART III

DISCUSSION

Chapter 10

EVALUATION OF THE MODEL

10.1 Predictions of the EVR to Natural Mating Calls

The proposed model relating the evoked vocal response to the acoustic stimulus has satisfactorily accounted for the evoked responses to the synthetic mating calls. The model has also been tested with natural mating calls and has satisfactorily predicted the vocal response to these calls as well.

Vocal responses have been evoked by the recorded bullfrog mating calls shown in Figures 5.1 to 5.4. The signal characteristics of these three mating calls satisfy the requirements of the model. In each call, a spectral peak occurs below 500 cps. Components in the high-frequency region (H) are simultaneously present, although their relative amplitudes are widely different in each call (see Chapter 5). These amplitude variations, however, are in keeping with the relative independence permitted for the amplitudes of the spectral components in region H and are within the required limits of the model. The temporal periodicities within the quasi-periodic croaks of these calls are approximately 100/sec. The model proposes that this is the most effective periodicity for evoking the vocal response.

For these three mating calls, vocal response to the laboratory call (Figures 5.1 and 5.4a) was less than to either the call obtained from Bogert (Figures 5.2 and 5.4b) or the call obtained from the Cornell Library of Natural Sounds (Figures 5.3 and 5.4c). In experimental sessions similar to those involving the synthetic mating calls, we obtained evoked-calling levels of 50 to 60% for the laboratory call and 75 to 90% for the field-recorded calls (64 to 76 db SPL). The predictions of the model are in agreement with these results. In each of the field-recorded calls, a low-frequency spectral peak occurs at approximately 200 cps. This is the proposed optimum location (i.e., region L) for evoking the vocal response. In the laboratory call, the low-frequency peak occurs at about 400 cps. The model predicts that, because of the suppressive effect of energy in region M, the degree of evoked calling will monotonically decrease as this peak is increased above 200 cps and, therefore, successfully predicts that vocal response should be less to the laboratory call.

The model was also tested by filtering these three mating calls. When these calls were sharply (102 db/oct) low-pass filtered at 500 cps or high-pass filtered at 1000 cps, vocal responses could not be evoked. The model's requirement of concomitant low-region and high-region energy predicts this result.

Not all bullfrog mating calls could evoke a calling response from our laboratory animals. In particular, calls recorded from some of the smaller, younger males in the New Jersey colony failed to evoke the vocal response. The gross temporal patterns of these calls were similar to those found in most other bullfrog mating calls (e.g., Figures 5.1a, 5.2a, and 5.3a). The waveform periodicities within the croaks of these calls tended toward slightly higher rates of 100 to 110 cps (compared to the periodicities of 95 to 100 cps usually found within the croaks made by the larger males). However, these slightly higher rates are well within the acceptable limits of the model.

Figure 10.1 shows representative spectrographic sections for two different mating calls recorded from two of the smaller males in the New Jersey colony. These sections, typical of the mating calls produced by the smaller males, each show a spectral peak in the region of 500 to 600 cps. The location of this peak varied inversely with body size: peak locations as high as 700 cps have been measured in calls made by the smallest males ($3\frac{1}{2}$ to 4 inches), whereas the mating calls of the larger males (6 to 7 inches) generally showed a spectral peak below 500 cps. The larger vocal-sac cavities of the older males are most likely responsible for this downward shift in lower-frequency energy.

Neither of the calls corresponding to the sections shown in Figure 10.1 was answered by the males of the laboratory colonies (despite the fact that one of the colonies included the same two smaller males that had made these calls). The failure of these calls to evoke a response agrees with the predictions of the model.

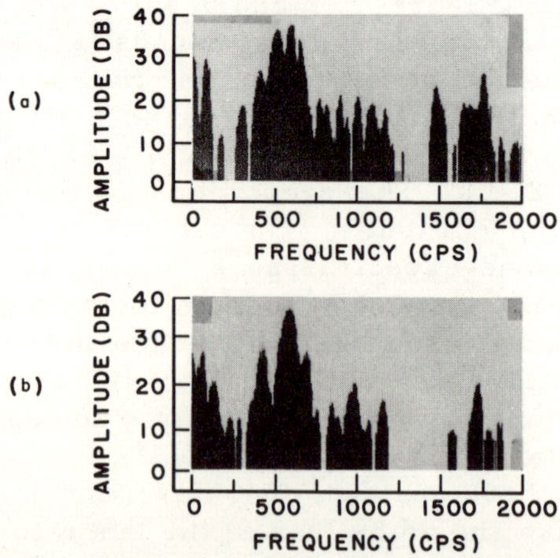

Figure 10.1. Two spectrographic sections representative of the croaks in two different mating calls recorded from two of the smaller male bullfrogs.

Suppression of calling is predicted to those sounds in which spectral amplitudes in M exceed those in L.

Vocal responses to other bullfrog mating calls have also been successfully predicted by the model. Since most of these calls showed similar waveform periodicities and had adequate energy distributions in the high-region, the spectral amplitudes in the low-region and mid-region were again critical in allowing prediction of the calling levels.

10.2 Electrophysiological Correlates

The proposed model represents an abstraction of behavioral results which reflect, in part, the auditory capabilities of the bullfrog. These capabilities depend, to a large degree, on the manner in which acoustic stimuli are represented in the auditory nervous system of this animal. Therefore, it should not be surprising that there is a close relationship between the behavioral results and the coding of acoustic stimuli in the bullfrog's nervous system. It is our belief that the selective behavioral response, evoked calling, is strongly predetermined at the peripheral level of the auditory nervous system. This belief is supported by the recent electrophysiological findings of Frishkopf and Goldstein.[63] Although these findings were briefly described in Chapter 4, they are again presented so as to stress their relevance to our behavioral results.

Frishkopf and Goldstein have recorded single-unit activity in the eighth nerve of anesthetized bullfrogs in response to acoustic

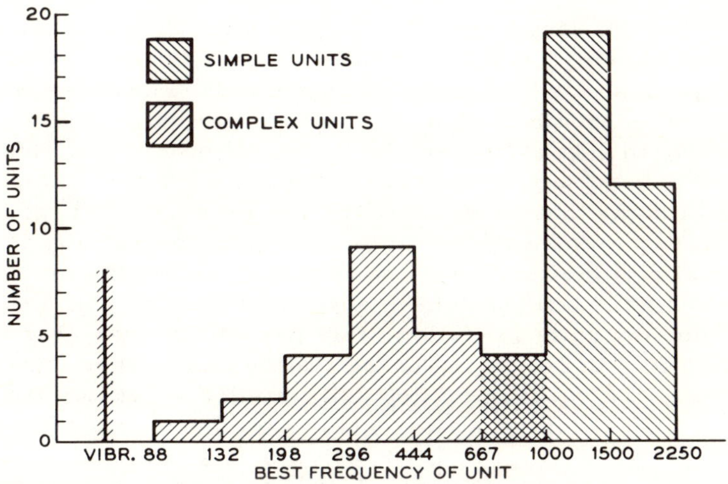

Figure 10.2. Histogram showing the best frequencies of the single units studied by Frishkopf and Goldstein (J. Acoust. Soc. Am. 35, 1223, 1963).

stimuli. On the basis of frequency sensitivity to tone-burst stimulation, they identified two distinct types of primary units. "Simple" units were most sensitive to frequencies between 1000 and 2000 cps; "complex" units were most sensitive to frequencies below 700 cps. (Within the complex class were units that also responded to vibration.) The histogram of Figure 10.2 shows the distribution of the best frequencies (i.e., most sensitive frequencies) of the units studied. The histogram, definitely bimodal, emphasizes the disjoint frequency sensitivities of the simple and complex types. Frishkopf and Geisler[79] later confirmed that these two types of units derive from separate papillae within the inner ear of the bullfrog: simple units from the basilar papilla and complex units from the amphibian papilla.

Simple and complex units could further be distinguished by Frishkopf and Goldstein. The responses of all complex units to an excitatory tone could be inhibited by the addition of a second tone which, when presented alone, had no effect. Inhibition was graded and most effective when the frequency of the inhibitory tone occurred in the neighborhood of 500 cps (regardless of the most sensitive excitatory frequency for that unit). For total inhibition, the level of the inhibitory tone had to exceed that of the excitatory tone (at the best frequency of that unit) by at least 6 db. This amplitude ratio for total inhibition was relatively independent of the acoustic level of the excitatory stimulus. In remarkable contrast, simple units could not be inhibited by the introduction of a second tone.

The ability of simple and complex units to code temporal fine structure was also investigated by Frishkopf and Goldstein. They found that the responses of both types were time-locked to trains of filtered pulses at pulse rates of 50 and 100/sec; at 200/sec, this synchronous "following" disappeared. For these pulse trains, the average number of neural spikes per unit of time corresponded, in decreasing order, to repetition rates of 100, 200, and 50 pulses/sec, respectively (for constant pulse amplitudes).

The finding that neural activity was greatest for repetition rates of 100/sec seems highly significant. This periodicity is characteristic of the temporal fine structure within the male's croak. Frishkopf and Goldstein have suggested that the ability of simple and complex units to "follow" this periodicity may play an important role in the coding of the mating call. They found that the playback of a bullfrog mating call initiated repetitive firing in both types of units at a rate equal to the waveform periodicity within the call. Figure 10.3 shows such a response from a complex unit. Note that the neural spikes, time-locked to the pulse structure within the croaks of the call, demonstrate an ability to code the temporal periodicity in the croak waveform.

The correlation between these electrophysiological findings and our evoked-calling results is quite striking. In fact, the

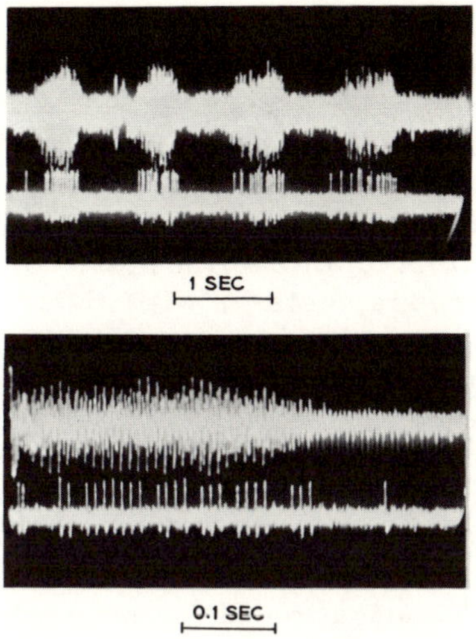

Figure 10. 3. Responses of a complex unit to a bullfrog mating call. Upper photograph: The playback of a mating call is represented in the upper trace, and the unit's corresponding response is shown in the lower trace. Lower photograph: Expanded display of the unit's response to a croak in the mating call (Frishkopf and Goldstein, J. Acoust. Soc. Am. 35, 1227, 1963).

correlation is so great as to suggest that the peripheral auditory system provides the neural basis for a major portion of the model. In order to evoke the vocal response, the stimulus requirement of concomitant energies in regions L and H indicates the necessity of a simultaneous excitation of the nerve fibers innervating the amphibian and basilar papillae. That the two papillae are anatomically distinct and that their corresponding single-unit populations are disjoint can be identified with the model's relative independence of the amplitudes of the low-region and high-region spectral envelopes. The finding that calling decreased as the amplitudes of the mid-region components were increased above the low-region spectral peak is consistent with the inhibition of complex units. Even stronger is the coincidental evidence that calling to the synthetic mating croaks could be suppressed by the introduction of a single 500-cps spectral line and that this same frequency produces greatest inhibition of the response of each complex unit to an excitatory stimulus. Furthermore, the relative amplitude of the 500-cps component (relative to the amplitude of the low-frequency spectral peak) required to suppress calling is in close agreement with the relative amplitude of the 500-cps component (relative to the amplitude of an excitatory stimulus) required to inhibit totally the responses of the complex units. Finally, that the amount of spike activity for both simple and complex units was greatest to pulse trains of 100 pulses/sec is consistent with the optimum waveform periodicity proposed by the model.

There remains one significant feature of the model which cannot be associated with the peripheral auditory system. The model requires that the neural responses arising from the low-region stimulation of the amphibian papilla must converge, somewhere in the bullfrog's nervous system, with those arising from the high-region stimulation of the basilar papilla. As far as is known, the responses of the complex and simple units are independent: there appears to be no interaction between cells of the two types. The convergence of low-region and high-region information must therefore be assigned to a higher nervous center. One might expect this interaction to occur within the central auditory system — but it could also occur within the efferent system involved in vocalization. An investigation of the neurophysiological mechanism of this convergence might well reveal the neural location of the true "mating-call detector" in this animal.

Chapter 11

SUGGESTIONS FOR FURTHER RESEARCH

This research has suggested many additional studies, both with the bullfrog as well as with other anuran species. An enumeration of all of these studies is impossible. Therefore, we shall discuss only a few of those experimental problems which we consider to relate most directly to questions of sound communication in the anuran.

11.1 Further Studies with the Bullfrog

Although our studies using synthesized mating croaks have stressed spectral structure and waveform periodicity, we should not assume that other properties of these stimuli were not essential in evoking the vocal response. Further experimental sessions involving such temporal parameters as croak fine structure (other than periodicity), croak envelope, silent interval between croaks, and number of croaks per call would expand the scope of our proposed model. Hopefully, such an expansion would maintain the predictive nature of the model and, therefore, could be tested with both synthetic and natural bullfrog mating calls. A word of caution is in order, however, in the contemplation of such further research. Sound reflections from laboratory surfaces can introduce serious distortions in the temporal structure of acoustic stimuli. Therefore, investigations of temporal parameters must be conducted in a suitable acoustic environment. Ideally, the laboratory should be soundproof and anechoic; the walls of the terrariums should be of screen or other acoustically transparent material. Since the bullfrog is very sensitive to frequencies as low as 100 to 200 cps, these ideal acoustics will be difficult to achieve – a laboratory that is anechoic at such low frequencies cannot be easily realized.

The proposed model applies strictly to the auditory capabilities of the male. An extension of these capabilities to the female remains to be made. Inasmuch as she did not vocally respond to either the natural or the synthetic mating calls, such an extension must involve a different behavioral response. One possibility would be a study of the approach response (see Chapters 1 and 4) of gravid female bullfrogs to synthetic mating calls. Rather than a seasonal restriction to ripe females captured in nature, the mating cycle could be initiated in laboratory females by hormonal injection. Extracts from the pituitary gland have induced metamorphosis in tadpoles[114,115] and ovulation in adult anuran females.[116,117]

In maintaining our colonies of laboratory bullfrogs, we found that occasionally the animals (both males and females) produced distinct vocalizations other than the mating call. These other types of calls have been recorded and are briefly described in Appendix A. Their biological significances and the bullfrog's ability to discriminate between them are each major studies in themselves and have not been investigated in detail. The acoustic presentation of these calls or their synthetic approximations may be capable of triggering different instinctive responses. For example, in our preliminary studies of the evoked vocal response, we discovered that the presentation of a single tone between 600 and 2500 cps would not initiate a vocal response but, instead, would produce a fear reaction: the animals (both males and females) immediately jumped into the terrarium ponds. That is, prior to the presentation of such a tone, only a very few animals were in the terrarium ponds at any given time — most of the animals maintained favorite terrestrial territories. As soon as the tonal presentation began, the animals left their territories and "escaped" into the water. Within 15 to 20 seconds, most of them had already entered the ponds. Upon termination of the tone, the animals slowly crawled out of the water and resumed their territories. In marked contrast, tones of approximately 400 cps or lower did not produce a fear reaction: the animals remained in their territories. The nature of this behavior is not understood nor was it further investigated. However, one particular type of call, the distress or pain cry of the bullfrog, is distinctly characterized by a high pitch. The association of this type of call with the escape response to high-frequency tones is an interesting conjecture (although Yerkes[40] has suggested that the distress call, by itself, does not trigger escape reactions in the anuran — see Chapter 2).

11.2 Studies with Other Species

The evoked vocal response is not unique to the bullfrog. We believe that a similar response can be evoked from the males of other species and that a study such as this can be conducted with other anuran species. We have tested this supposition in a preliminary study with a laboratory colony of green frogs (*Rana clamitans*). The colony (22 females, 26 males) was originally captured in Wisconsin and has been maintained in a terrarium under the same conditions as those for the laboratory bullfrogs (Chapter 6). The results of our preliminary study with the green frogs are now summarized.

11.2.1 Mating Call of the Green Frog.

In contrast to the series of croaks within the mating call of the bullfrog, the green

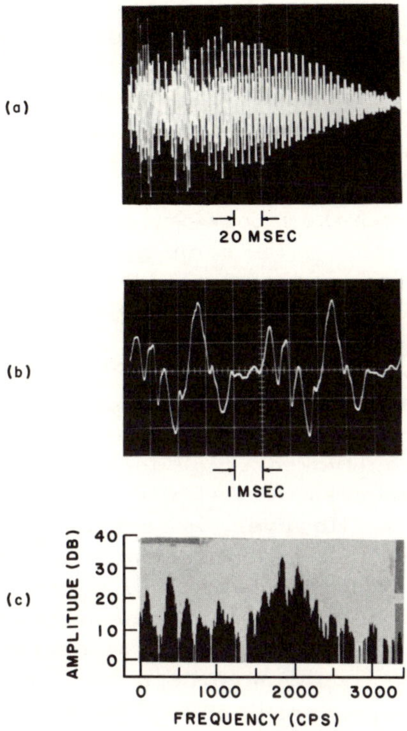

Figure 11.1. Field-recorded green frog mating call. This recording was obtained from W. F. Blair (Department of Zoology, University of Texas, Austin, Texas). (a) The mating call in its entirety. (b) Expansion of the waveform in the body of the call. (c) A representative spectrographic section through the body of the call.

frog's mating call consists of but a single croak — an explosive, banjo-like sound. Figure 11.1 shows a representative mating call of the green frog.* This call, made only by the male, is produced by a rapid inflation of his paired lateral vocal sacs. The temporal envelope (Figure 11.1a) of the call shows a sudden onset, followed by a gradual decay. The call generally lasts for about 0.2 sec. The temporal fine structure (Figure 11.1b) within the body of the call has a repetition rate of approximately 200/sec. Spectrographic sections (Figure 11.1c), taken through the middle of the call, usually show a concentration of low-frequency energy

*Recordings of the mating call of the green frog were obtained from W. F. Blair (Department of Zoology, University of Texas, Austin, Texas) and from the Cornell Library of Natural Sounds (Laboratory of Ornithology, Cornell University, Ithaca, New York).

centered about 400 to 500 cps and a broad distribution of high-frequency energy over the range of 1500 to 2500 cps.

11.2.2 The EVR of the Green Frog to Natural Mating Calls. The selected excerpts (Chapter 7) of the mating calls of the green frog, the bullfrog, the southern bullfrog (*Rana grylio*), and the meadow frog (*R. p. pipiens*) were tested for their ability to evoke a calling response from the laboratory colony of green frogs. This study was conducted in the same manner as the earlier study of the evoked vocal response of the bullfrog to natural mating calls (see Chapter 7). The excerpts of the four species were presented in random order at a trial rate of one per hour; all presentations were at approximately 70 db SPL.

Whereas the excerpts of the bullfrog, southern bullfrog, and meadow frog were not answered by the laboratory colony, the excerpt of the green frog evoked a high degree of calling. Most of the evoked vocal responses were similar to the green frog's mating call (Figure 11.1). However, in several instances, vocal responses could not be readily identified with the mating call. No attempt was made to classify these other vocalizations into different types. They may have been due to immature males attempting a mating call. (Some of the males had a body length of only 2 inches.) They also could have been made by females. In any case, vocal responses, regardless of their type, were evoked only by the mating calls of the green frog. This selectivity closely parallels that observed in our studies with the bullfrog and, again, indicates the species specificity of the anuran mating call.

11.2.3 The EVR of the Green Frog to Filtered Green Frog Mating Calls. A study of the vocal responses of the laboratory green frogs to filtered green frog mating calls was also conducted. Hourly trials, presented in random sequence, involved low-pass filtered (102 db/oct), high-pass filtered (102 db/oct), and unfiltered playbacks of an excerpt of a green frog mating chorus. The excerpt lasted for 150 seconds and was taken from a recording made in Texas during the green frog's breeding season (obtained from W. F. Blair, University of Texas). The signal characteristics of all of the mating calls in this excerpt were very similar to those of Figure 11.1. All trials were presented at approximately 70 db SPL. A measure of the relative amount of calling for different filter settings was obtained by counting the number of individual vocal responses, regardless of type, evoked during each playback. For each filter setting, an average number of responses (based on 10 trials) was computed and then expressed as a percentage of the average number of responses evoked by the unfiltered (control) playbacks. The unfiltered excerpt, presented on every fifth trial, was used as an experimental control: if its presentation failed to evoke a high degree of calling, then that experimental session was terminated and trials were not resumed until the following day.

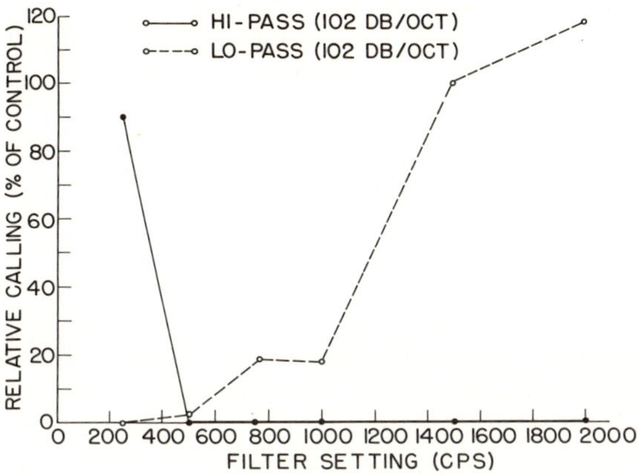

Figure 11.2. Relative amount of calling evoked by the filtered playbacks of a green frog mating chorus. The amount of calling is expressed as a percentage of the average number of responses evoked by the unfiltered playback (control).

The results of this preliminary study are shown in Figure 11.2. The most outstanding feature of these data is the disjunctive nature, for high calling levels, of the low-pass and high-pass filter settings. This study suggests that, to evoke the vocal response of the green frog, sufficient energy must be simultaneously present below 500 cps and above 1000 cps. A comparison with the evoked vocal response of the bullfrog shows a remarkable similarity. For both the bullfrog and the green frog, the vocal response is dependent upon concomitant low-frequency and high-frequency energy. The comparison is even more remarkable in view of the obvious differences in the temporal patterns of the mating calls of the two species.

11.2.4 Electrophysiological Correlates in the Green Frog. Sachs,[80] as mentioned in Chapter 4, has recorded from single units in the eighth nerve of the green frog in response to acoustic stimuli. Based on frequency sensitivity, two disjoint populations were found. Units of one population were most sensitive to tones above 650 cps. These (high-frequency) units could not be inhibited by the introduction of a second tone. Units of the other population were most sensitive to tones below 450 cps. All of these (low-frequency) units could be inhibited by the introduction of a second tone; 700 cps was the most effective frequency for inhibition of these units (regardless of their most sensitive frequency for excitation).

The fit of these electrophysiological findings to our preliminary studies of the evoked vocal response of the green frog is highly suggestive. It would appear that the evoked vocal response has its origin in the simultaneous excitation of the high-frequency and low-frequency units found by Sachs. Therefore, based on our studies with the bullfrog, we might suspect that the peripheral auditory system of the green frog plays an important role in the coding of the mating call. We might further suspect that the evoked vocal response of the green frog can be suppressed by a sufficient increase in the relative amplitude of mid-range energy in the neighborhood of 700 cps (i.e., the most effective frequency for inhibition of low-frequency units). This was not investigated in our preliminary studies with the green frog.

11.2.5 Concluding Remarks. Our studies of the evoked vocal responses of the bullfrog and the green frog have indicated a close relationship to the coding of acoustic stimuli in the nervous systems of these two species. In fact, this relationship is so close as to suggest that, within the frog, one finds an excellent opportunity to study simultaneously a biologically significant behavioral response and its neurophysiological representation. We hope that studies of the evoked vocal response can be extended to other species. Such studies, when combined with appropriate neurophysiological studies, may lead us to a better understanding of sound communication in the anuran.

Appendix A

CALLS RECORDED FROM LABORATORY BULLFROGS

In the course of this research, we noted that the bullfrogs of our laboratory colonies often produced vocalizations other than the mating call. These vocalizations could be stereotyped into several distinct calls, each of which seemed to be associated with a different class of behavior. We have recorded these calls and have made a very preliminary analysis of their signal characteristics.

A.1 Mating Call

The bullfrog's mating call has been discussed in Chapter 5. Of the different types of calls that we have recorded from our laboratory bullfrogs, only the playback of this call was capable of evoking a vocal response.

A.2 Territorial Calls

These calls have been classified as territorial on the basis of the conditions under which they occurred. That is, many of the animals tended to maintain exclusive locations in the terrariums. When another bullfrog of that colony chanced to enter this location, a distinct call was often uttered by the proprietary frog (commonly followed with an attack upon the intruder – see Chapter 2). We found three different types of these calls. One type was made only by males, a second type was made only by females, and a third type was made by both males and females.

The male's call, shown in Figure A.1, resembles a single croak in the bullfrog's mating call (Figures 5.1 to 5.4). Its waveform, usually lasting from 0.4 to 0.6 sec, has a pulsatile fine structure with a repetition rate of approximately 100/sec. We found three characteristics of this call which reliably distinguished it from the mating call. First, whereas the mating call always consisted of a sequence of at least three or more repeated croaks, there was only one isolated vocalization (represented by the waveform in Figure A.1a) ever given in this call; i.e., it was never made repetitively. Second, the distribution of energy in the mating call of the adult male typically showed a low-frequency spectral peak, located below 400 cps, whose amplitude clearly exceeded that of the components between 500 and 700 cps. In comparison, the territorial call usually had a spectral peak in the neighborhood of 500 to 700 cps with an amplitude equal to or greater than that of

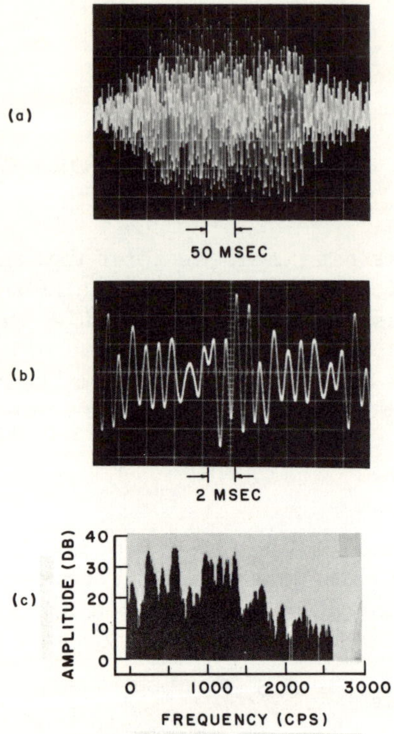

Figure A.1. Territorial call made only by the males of the laboratory colonies of bullfrogs. (a) The call in its entirety. (b) Expansion of the call, showing the repetitive structure of the waveform (approximately 100/sec). (c) A representative spectrographic section taken through the middle of the call.

the lower-frequency components. And third, the vocal sacs of the male were also involved in the production of the territorial call. However, in making this call, it appeared to us that he only partially inflated his vocal sacs (as compared with the full expansion, quite noticeable, in his production of the mating call). Such a partial expansion of the vocal sacs could account for the increased presence of energy between 500 and 700 cps.

The female's territorial call is shown in Figure A.2. This call, consisting of a single vocalization, generally had a duration of 1.5 to 2.0 sec. Its quasi-periodic waveform, quite pulsatile, usually showed a repetition rate of 25 to 30/sec. This call, as was the case for the male's call, was most often made during feeding periods wherein the members of the colony commonly trespassed individual territories in search of food.

A third type of call, made equally well by both sexes, is shown in Figure A.3. This call was often given by a proprietary frog

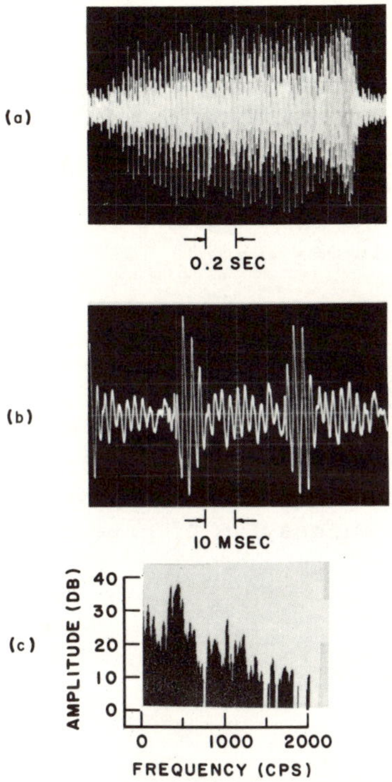

Figure A.2. Territorial call made only by the females of the laboratory colonies of bullfrogs. (a) The call in its entirety. (b) Expansion of the call, showing the repetitive structure of the waveform (approximately 28/sec). (c) A spectrographic section taken through the pulsed portion of an individual period in the waveform.

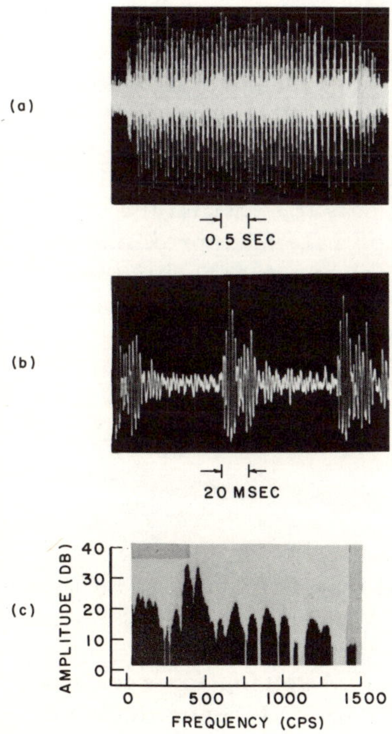

Figure A.3. Territorial call made by both males and females of the laboratory colonies of bullfrogs. (a) The call in its entirety. (b) Expansion of the call, showing the repetitive structure of the waveform (approximately 12/sec). (c) A spectrographic section taken through the pulsed portion of an individual period in the waveform.

immediately after the successful banishment of an intruder. It also was sometimes made shortly after an extraneous disturbance in the laboratory (e.g., changing the water in the ponds, sudden noises). This call was uniquely characterized by a long duration (commonly from 3 to 6 sec) and a quasi-periodic waveform with a very low fundamental frequency (usually between 10 and 15 cps). The temporal fine structure in the waveform is highly pulsatile, giving rise to a sound similar to a train of clicks.

A. 3 Release Call

The males, in the course of maintaining territories and competing for food, would often clasp another member (males and females alike) of their colony. The male usually maintained his hold, whereupon the clasped animal emitted a characteristic call. (See Chapter 2 for the significance of this behavior.) Figure A. 4 shows such a call. It consisted of a succession of repeated

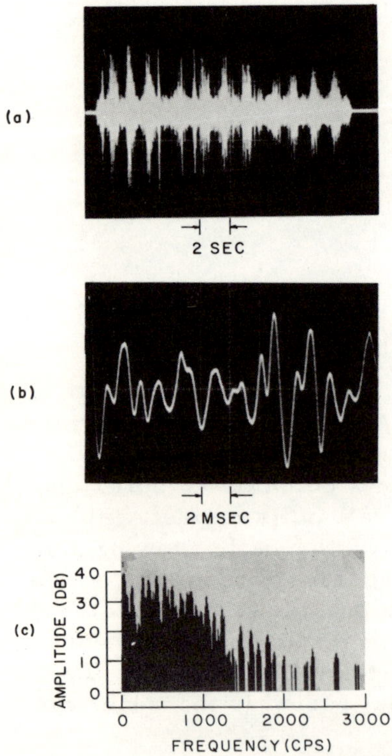

Figure A. 4. Release call made by both males and females of the laboratory colonies of bullfrogs. (a) A sample release call shown in its entirety. (b) Expansion of one of the repeated sounds in the call. The waveform generally showed a pulsatile structure with a repetition rate of approximately 75/sec. (c) Spectrographic section corresponding to the above waveform.

sounds which continued until the animal was released. The gross temporal properties of this call varied widely, probably because of the considerable thrashing about the terrarium that accompanied the clasping of an unwilling animal. Allowing for such variations, we were unable to note a significant difference between the call of the male and that of the female. Generally, the waveform showed a pulsatile structure having a repetition rate (usually) between 70 and 85/sec.

A.4 Warning Call

The males and females of our bullfrog colonies often produced a call sounding very much like a short, loud "grunt." Naturalists, upon observing instances of this call in several other species, have proposed that it provides a warning function. However, as we discussed in Chapter 2, our observations have suggested that this call might serve simply to deflate the lungs, in which case a warning significance seems questionable.

A typical call from one of our laboratory animals is shown in Figure A.5. The most characteristic features of this sound are its sudden onset and brief duration (generally lasting about 0.1 sec). Whereas the waveforms of the other bullfrog calls usually had a repetitive structure, the temporal fine structure within

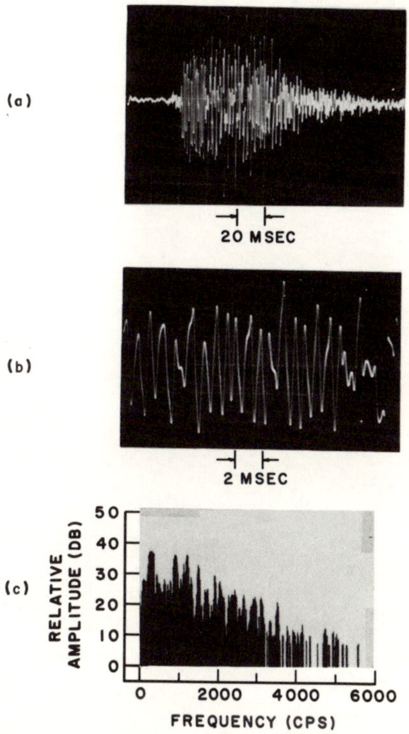

Figure A.5. Warning call made by both males and females of the laboratory colonies of bullfrogs. (a) The call in its entirety. (b) Expansion of the waveform. (c) Spectrographic section taken through the middle of the call.

this call varied widely (for different animals as well as for the same animal at different times). The spectral envelope of its energy distribution often showed a gradual decrease in amplitude with increasing frequency. We have been unable to observe any significant differences between the male's call and the female's call.

A.5 Distress Call

We have observed the bullfrog's distress call made in a few instances by our laboratory animals. Unfortunately, we were unable to record this sound. It has been made by both males and females. The most distinctive features of this call are its high pitch and its mode of production (made with the mouth wide open).

Appendix B

TREATMENT OF DISEASED BULLFROGS

Most anurans, particularly the more aquatic species, are quite susceptible to a number of fatal, highly contagious diseases. Inasmuch as these diseases are poorly understood, their treatment and cure are a matter of experimentation. As was mentioned in Chapter 6, we have occasionally encountered diseases among our laboratory bullfrogs. In attempting to cure these diseased animals, a routine treatment was adopted which, in a majority of cases, has led to a successful recovery. Since this treatment has been relatively successful, it will be briefly described.

We employed the same treatment for all diseases. The first stage of this treatment involved a prolonged exposure to a constant temperature of 37 °F: the animal, placed in a large container, was maintained under refrigeration for a period of several weeks. The container was filled to a depth of 1 to 2 inches with tap water in which a small amount of Acriflavine (Chapter 6) had been added for antiseptic purposes. This solution was usually changed every other day; a reserve of tap water, also maintained at 37 °F, was used for refill. The animal was not fed during the entire refrigeration period. (At these low temperatures, the animals remain in a quasi-hibernating state and can endure prolonged durations without food.)

The exposure to the low temperature, while it did not cure the animal, did arrest the spread of the disease: his condition remained about the same throughout this period. After approximately 4 to 6 weeks under refrigeration, the animal was transferred to an isolated terrarium for continued treatment and cure (hopefully). The terrarium, similar to those previously described (Chapter 6), was kept especially for sick animals. The environmental temperature, as for all of our terrarium animals, was maintained at 67 to 72 °F. The water in the terrarium pond was refilled every second day with fresh tap water (temperature between 60 and 70 °F). To this water were added a small amount of Acriflavine (Chapter 6) and several tablets* of Terramycin (with vitamins A, D, and Niacinamide). The concentration of Terramycin was approximately 100 mg per gallon of water.

---

*The Terramycin was obtained in the form of scours tablets from C. Pfizer and Co., Brooklyn, New York.

The antibiotic effects of the Terramycin produced a distinct improvement in the condition of our diseased animals (in most cases). Usually signs of recovery were evident within 4 to 7 days. This treatment was continued until the animal showed normal, active behavior and all signs of his disease had disappeared. During this entire period, the animals were fed live insects (mealworms, crickets) every 2 or 3 days. Upon a successful recovery, the animal was then returned to his original terrarium colony. If he showed any signs of relapse during his recovery, he was immediately placed under refrigeration; the entire treatment was then repeated.

REFERENCES

1. R. A. Mc Cleary and J. J. Bernstein, "A unique method for control of brightness cues in study of color vision in fish," Physiol. Zool. 32, 284-292 (1959).

2. J. J. Bernstein, "Loss of hue discrimination in forebrain-ablated fish," Exptl. Neurol. 3, 1-17 (1961).

3. J. J. Bernstein, "Brightness discrimination following forebrain ablation in fish," Exptl. Neurol. 3, 297-306 (1961).

4. W. G. Reese and R. A. Dykman, "Conditional cardiovascular reflexes in dogs and men," Physiol. Rev. 40 (Supp. No. 4), 250-265 (1960).

5. W. H. Gantt, "Cardiovascular component of the conditional reflex to pain, food and other stimuli," Physiol. Rev. 40 (Supp. No. 4), 266-291 (1960).

6. S. Soltysik, K. Jaworska, M. Kowalska, and S. Radom, "Cardiac responses to simple acoustic stimuli in dogs," Acta Biol. Exptl. 21, 235-252 (1961).

7. J. E. Peters and W. H. Gantt, "Conditioning of human heart rate to graded degrees of muscular tension," Federation Proc. 10, p. 104 (1951).

8. R. A. Dykman and W. H. Gantt, "Autonomic conditioning in dogs and humans," Physiological Bases of Psychiatry, edited by W. H. Gantt (Charles C. Thomas, Publisher; Springfield, Illinois, 1958), pp. 171-195.

9. R. A. Dykman, W. G. Reese, C. R. Galbrecht, and P. J. Thomasson, "Psychophysiological reactions to novel stimuli: measurement, adaptation, and relationship of psychological and physiological variables," Ann. N. Y. Acad. Sci. 79, p. 43 (1959).

10. B. M. Shmavonian, "Methodological study of vasomotor conditioning in human subjects," J. Comp. Physiol. Psychol. 52, 315-321 (1959).

11. B. S. Martof and E. F. Thompson, "Reproductive behavior of the chorus frog (Pseudacris nigrita)," Behaviour 13, 243-258 (1958).

12. B. S. Martof, "Vocalization as an isolating mechanism in frogs," Am. Midland Naturalist 65, 118-126 (1961).

13. W. F. Blair and M. J. Littlejohn, "Stage of speciation of two allopatric populations of chorus frogs (Pseudacris)," Evolution 14, 82-87 (1960).

14. M. J. Littlejohn, "Call discrimination and potential reproductive isolation in Pseudacris triseriata females from Oklahoma," Copeia 4, 370-371 (1960).

15. M. J. Littlejohn, "Mating call discrimination by females of the spotted chorus frog (Pseudacris clarki)," Texas J. Sci. 13, 49-50 (1961).

16. M. J. Littlejohn and T. C. Michaud, "Mating call discrimination by females of Strecker's chorus frog (Pseudacris streckeri)," Texas J. Sci. 11, 86-92 (1959).

17. M. J. Littlejohn, M. J. Fouquette, and C. Johnson, "Call discrimination by female frogs of the Hyla versicolor complex," Copeia 1, 47-49 (1960).

18. S. J. Holmes, The Biology of the Frog (The Macmillan Co., New York, 1934), Fourth Revised Edition, pp. 2-25.

19. G. K. Noble, The Biology of the Amphibia (Dover Publications, Inc., New York, 1954), pp. 459-540.

20. Ibid., pp. 1-13.

21. A. H. Wright and A. A. Wright, Handbook of Frogs and Toads of the United States and Canada (Comstock Publishing Associates, Ithaca, New York, 1949), p. 2.

22. D. M. Cochran, Living Amphibians of the World (Doubleday and Co., Inc., Garden City, N. Y., 1961), pp. 15-17.

23. Ibid., pp. 17-46.

24. S. Ferhat-Akat, "Untersuchungen über den Gehörsinn der Amphiben," Z. Vergleich. Physiol. 26, 253-281 (1938).

25. G. K. Noble, op. cit., pp. 341-343, 356-369 (Ref. 19).

26. D. M. Cochran, op. cit., pp. 46-176 (Ref. 22).

27. C. M. Bogert, "The influence of sound on the behavior of amphibians and reptiles," Animal Sounds and Communication, edited by W. E. Lanyon and W. N. Tavolga (American Institute of Biological Sciences, Washington, D.C., 1960), pp. 137-320.

28. C. C. Liu, "Types of vocal sac in the Salientia," Proc. Boston Soc. Nat. Hist. 41, 19-40 (1935).

29. A. H. Wright and A. A. Wright, op. cit., passim (Ref. 21).

30. D. M. Cochran, op. cit., passim (Ref. 22).

31. H. R. Smyth, Amphibians and Their Ways (The Macmillan Co., New York, 1962), pp. 243-252.

32. V. K. Markuze, "Sezonnoe izmenenie vstrechaemosti samtsov i samok ozernoi lyagushki (Rana radibunda)," Nauchn. Dokl. Vysshei. Shkoly 3, 48-50 (1961).

33. T. Koppanyi and J. F. Pearcy, "Studies on the clasping reflex in Amphibia," Am. J. Physiol. 71, 34-39 (1924).

34. S. J. Holmes, op. cit., pp. 55-56 (Ref. 18).

35. H. R. Smyth, op. cit., p. 9-25 (Ref. 31).

36. G. K. Noble and E. J. Farris, "The method of sex recognition in the woodfrog (Rana sylvatica Le Conte)," Am. Museum Novitates 363, 1-17 (1929).

37. G. K. Noble, op. cit., pp. 406-412 (Ref. 19).

38. S. J. Holmes, op. cit., pp. 53-54 (Ref. 18).

39. F. H. Test, "Social aggressiveness in an amphibian," Science 120, 140-141 (1954).

40. R. M. Yerkes, "The instincts, habits, and reactions of the frog," Psychol. Rev. Monog. 4, 579-638 (1903).

41. W. F. Blair, "Call difference as an isolation mechanism in Florida species of hylid frogs," Quart. J. Florida Acad. Sci. 21, 32-48 (1958).

42. W. F. Blair, "Mating call in the speciation of anuran amphibians," Am. Naturalist 92, 27-51 (1958).

43. C. H. Lowe, Jr., "Isolating mechanisms in sympatric populations of southwestern anurans," Texas J. Sci. 6, 265-270 (1954).

44. P. P. Kellogg, "Considerations and techniques in recording sound for bio-acoustics studies," Animal Sounds and Communication, edited by W. E. Lanyon and W. N. Tavolga (American Institute of Biological Sciences, Washington, D. C., 1960), pp. 1-25.

45. M. J. Littlejohn, "Call differentiation in a complex of seven species of Crinia (Anura, Leptodactylidae)," Evolution 13, 452-468 (1959).

46. M. J. Littlejohn and A. R. Main, "Call structure in two genera of Australian burrowing frogs," Copeia 3, 266-270 (1959).

47. W. F. Blair, "Call difference as an isolation mechanism in southwestern toads (genus Bufo)," Texas J. Sci. 8, 87-106 (1956).

48. W. F. Blair, "Mating call and stage of speciation in the Microhyla olivacea – M. carolinensis complex," Evolution 9, 469-480 (1955).

49. W. F. Blair, "Differentiation of mating call in spadefoots, genus Scaphiopus," Texas J. Sci. 7, 183-188 (1955).

50. E. F. Thompson and B. S. Martof, "A comparison of the physical characteristics of frog calls (Psuedacris)," Physiol. Zool. 30, 328-341 (1957).

51. G. B. Rabb, "On the unique sound production of the Surinam toad (Pipa pipa)," Copeia 4, 368-369 (1960).

52. G. K. Noble, op. cit., pp. 87-88 (Ref. 19).

53. G. K. Noble and P. G. Putnam, "Observations on the life history of Ascaphus truei Stejneger," Copeia 3, 97-101 (1931).

54. R. R. Stuart, The Anatomy of the Bullfrog (Denoyer-Geppert Co., Chicago, 1958), p. 26.

55. S. J. Holmes, op. cit., pp. 168-175 (Ref. 18).

56. E. Trewavas, "The hyoid and larynx of the Anura," Phil. Trans. Roy. Soc. London B222, 401-527 (1933).

57. G. K. Noble, op. cit., pp. 158-176 (Ref. 19).

58. W. H. Mc Alister, "The vocal structure and method of call production in the genus Scaphiopus Holbrook," Texas J. Sci. 11, 60-77 (1959).

59. R. C. Zweifel, "Effects of temperature on calls of the frog Bombina variegata," Copeia 4, 322-327 (1959).

60. N. F. Patterson, "The inner ear of some members of the Pipidae (Amphibia)," Proc. Zool. Soc. London 134, 509-546 (1960).

61. A. S. Romer, The Vertebrate Body (W. B. Saunders Co., Philadelphia, 1960), Second Edition.

62. W. A. van Bergeijk and E. Witschi, "The basilar papilla of the anuran ear," Acta Anat. 30, 81-91 (1957).

63. L. S. Frishkopf and M. H. Goldstein, Jr., "Responses to acoustic stimuli from single units in the eighth nerve of the bullfrog," J. Acoust. Soc. Am. 35, 1219-1228 (1963).

64. C. D. Geisler, W. A. van Bergeijk, and L. S. Frishkopf, "The inner ear of the bullfrog," J. Morphol. 114, 43-58 (1964).

65. H. M. de Burlet, "Zur vergleichenden Anatomie der Labyrinthinnervation," J. Comp. Neurol. 47, 155-169 (1929).

66. O. Larsell, "The differentiation of the peripheral and central acoustic apparatus in the frog," J. Comp. Neurol. 60, 473-527 (1934).

67. C. U. A. Kappers, G. C. Huber, and E. C. Crosby, The Comparative Anatomy of the Nervous System of Vertebrates, Including Man (Hafner Publishing Co., New York, 1960), pp. 500-502, 944-946, 1196-1197.

68. J. W. Papez, Comparative Neurology (Hafner Publishing Co., New York, 1929), pp. 431-442, 500-503.

69. C. U. A. Kappers, G. C. Huber, and E. C. Crosby, op. cit., p. 1170 (Ref. 67).

70. W. E. Le Gros Clark, "The medial geniculate body and the nucleus isthmi," J. Anat. 67, 536-548 (1933).

71. G. von Békésy, Experiments in Hearing (McGraw-Hill Book Co., Inc., New York, 1960).

72. H. Davis, "Biophysics and physiology of the inner ear," Physiol. Rev. 37, 1-49 (1957).

73. R. Galambos, "Neural mechanisms of audition," Physiol. Rev. 34, 497-528 (1954).

74. W. A. van Bergeijk, "Observations on models of the basilar papilla of the frog's ear," J. Acoust. Soc. Am. 29, 1159-1162 (1957).

75. W. F. Strother, "The electrical response of the auditory mechanism in the bullfrog (Rana catesbeiana)," J. Comp. Physiol. Psychol. 52, 157-162 (1959).

76. E. D. Adrian, K. J. W. Craik, and R. S. Sturdy, "The electrical response of the auditory mechanism in cold-blooded vertebrates," Proc. Roy. Soc. London B125, 435-455 (1938).

77. G. V. Glekin and G. M. Erdman, "Discrimination of a useful signal by the auditory analyzer – I. Potentials from the elements of the frog auditory nerve," Biophysics 5, 474-481 (1960).

78. F. S. Axelrod, as reported by J. Y. Lettvin and H. R. Maturana, "Hearing senses in the frog," Quarterly Progress Report No. 57, Research Laboratory of Electronics, M.I.T., Cambridge, Mass. (1960) pp. 167-168.

79. L. S. Frishkopf and C. D. Geisler, in preparation, 1964.

80. M. B. Sachs, "Characteristics of primary auditory neurons in the green frog," S.M. Thesis, Department of Electrical Engineering, Mass. Inst. Tech., Cambridge, Mass., 1964.

81. R. M. Yerkes, "The sense of hearing in frogs," J. Comp. Neurol. Psychol. 15, 279-304 (1905).

82. G. Hinsche, "Kampfreaktionen bei einheimischen Anuren," Biol. Zentrbl. 48, 577-616 (1928).

83. G. Hinsche, "Untersuchungen über den Augenschlussreflex bei Bufo vulgaris und einige seiner Beziehungen zu anderen Reaktionen," Biol. Zentrbl. 46, 742-747 (1926).

84. R. Kuroda, "Experimental researches on the sense of hearing in lower vertebrates, including reptiles, amphibians, and fishes," Comp. Psychol. Monog. 3, 1-50 (1926).

85. C. Corbeille, "L'influence des vibrations acoustiques sur la respiration chez la grenouille et certains mammifères," C.R. Soc. Biol. 101, 113-115 (1929).

86. B. I. Bajandurow and W. A. Pegel, "Der bedingte Reflex bei Fröschen," Z. Vergleich Physiol. 18, 284-297 (1932).

87. E. R. Hilgard and D. G. Marquis, Conditioning and Learning, revised by G. A. Kimble (Appleton-Century-Crofts, Inc., New York, 1961).

88. F. Schwarz and E. Volkmer, "Eine sinnesphysiologische Methode zur Messung Hörverlustes nach Schallbelastung bei Fröschen," Acta Biol. Med. Ger. 8, 290-294 (1962).

89. T. E. Mc Gill, "A review of hearing in amphibians and reptiles," Psychol. Bull. 57, 165-168 (1960).

90. W. F. Strother, "Hearing in frogs," J. Aud. Res. 2, 279-286 (1962).

91. G. W. Goldsmith, "Habits and reactions of Scaphiopus hammondi," Yearbook Carnegie Inst. Wash. 25, 369-370 (1925-26).

92. M. J. Murray, "Analysis and synthesis of the mating croak of the bullfrog," S.B. Thesis, Department of Electrical Engineering, Mass. Inst. Tech., Cambridge, Mass., 1964.

93. L. G. Kersta, "Amplitude cross-section representation with the sound spectrograph," J. Acoust. Soc. Am. 20, 796-801 (1948).

94. M. V. Mathews, J. E. Miller, and E. E. David, Jr., "Pitch synchronous analysis of voiced sounds," J. Acoust Soc. Am. 33, 179-186 (1961).

95. N. R. French and J. C. Steinberg, "Factors governing the intelligibility of speech sounds," J. Acoust. Soc. Am. 19, 90-119 (1947).

96. R. K. Potter and G. E. Peterson, "The representation of vowels and their movements," J. Acoust. Soc. Am. 20, 528-535 (1948).

97. R. K. Potter and J. C. Steinberg, "Toward the specification of speech," J. Acoust. Soc. Am. 22, 807-820 (1950).

98. G. E. Peterson, "The information bearing elements of speech," J. Acoust. Soc. Am. 24, 629-637 (1952).

99. R. L. Miller, "Auditory tests with synthetic vowels," J. Acoust. Soc. Am. 25, 114-121 (1953).

100. H. K. Dunn, "The calculation of vowel resonances and an electrical vocal tract," J. Acoust. Soc. Am. 22, 740-753 (1950).

101. J. L. Flanagan, "Difference limen for vowel formant frequency," J. Acoust. Soc. Am. 27, 613-617 (1955).

102. J. L. Flanagan, "Note on the design of terminal-analog speech synthesizers," J. Acoust. Soc. Am. 29, 306-310 (1957).

103. R. L. Miller, "Nature of the vocal cord wave," J. Acoust. Soc. Am. 31, 667-677 (1959).

104. I. Lehiste and G. E. Peterson, "Some basic considerations in the analysis of intonation," J. Acoust. Soc. Am. 33, 419-425 (1961).

105. G. Fant, "Acoustic analysis and synthesis of speech with applications to Swedish," L. M. Ericsson Technics No. 1, Chapters 1-3 (1959).

106. E. S. Weibel, "Vowel synthesis by means of resonant circuits," J. Acoust. Soc. Am. 27, 858-865 (1955).

107. K. N. Stevens, R. P. Bastide, and C. P. Smith, "Electrical synthesizer of continuous speech," J. Acoust. Soc. Am. 27, p. 207 (1955), Abstract.

108. H. R. Smyth, op. cit., pp. 88-89 (Ref. 31).

109. G. K. Noble, op. cit., pp. 120-121 (Ref. 19).

110. Ibid. pp. 431-444.

111. S. J. Holmes, op. cit., pp. 43-48 (Ref. 18).

112. A. T. Cameron, "Further experiments on conditions influencing the life history of the frog," Proc. Trans. Roy. Soc. Can. 15 (Sec. V), 13-21 (1921).

113. R. W. Young, "Single-number criteria for room noise," J. Acoust. Soc. Am. 36, 289-295 (1964).

114. W. Etkin, "Metamorphosis-activating system of the frog," Science 139, 810-814 (1963).

115. H. Lipner and S. Hazen, "Extrathyroidal iodide pump in tadpoles (Rana grylio)," Science 138, 898-899 (1962).

116. R. Rugh, "Pituitary-induced sexual reactions in the anura," Biol. Bull. 68, 74-81 (1935).

117. O. M. Wolf, "Effects of daily transplants of anterior lobe of the pituitary on reproduction of the frog (Rana pipiens Shreber)," Anat. Record 44, p. 206 (1929).

INDEX

163